Cutting the Cost of Flying

Other books by Geza Szurovy

The Private Pilot's Guide to Renting & Flying Airplanes Worldwide, TAB/McGraw-Hill, Inc.

Fly for Less: Flying Clubs & Aircraft Partnerships, TAB/McGraw-Hill, Inc.

Basic Aerobatics with coauthor Mike Goulian, TAB/McGraw-Hill, Inc.

Cutting the Cost of Flying

Geza Szurovy

TAB Books
Division of McGraw-Hill, Inc.
New York San Francisco Washington, D.C. Auckland Bogotá
Caracas Lisbon London Madrid Mexico City Milan
Montreal New Delhi San Juan Singapore
Sydney Tokyo Toronto

Disclaimer:
All material in this book is intended to be used as general information only. The contents, including all forms and tables, are not to be relied upon as specific legal, financial, maintenance or aircraft operating advice. It is the reader's responsibility to seek the services of appropriate attorneys at law and insurance, financial, and maintenance professionals for advice specific to each individual reader's needs. It is the pilot in command's responsibility to consult all relevant sources of official information relevant to every aspect of a proposed flight, and personally assure compliance with all laws, regulations, and procedures.

Published by TAB Books, an imprint of McGraw-Hill, Inc.

1 2 3 4 5 6 7 8 9 0 DOH/DOH 9 9 8 7 6 5 4

Library of Congress Cataloging-in-Publication Data
Szurovy, Geza.
Cutting the cost of flying / by Geza Szurovy.
p. cm.
Includes index.
ISBN 0-07-062993-5 (pbk.)
1. Private flying—Costs. 2. Airplanes, Private—Cost of operation. I. Title.
TL721.4.S977 1994
629.132'5217'0681—dc20 93-50703
CIP

Acquisitions editor: Jeff Worsinger
Editorial team: Robert E. Ostrander, Executive Editor
Norval G. Kennedy, Book Editor
Production team: Katherine G. Brown, Director
Susan E. Hansford, Coding
Patsy D. Harne, Desktop Operator
Linda L. King, Proofreading
Stephanie Meyers, Computer Artist
Elizabeth J. Akers, Indexer
Design team: Jaclyn J. Boone, Designer
Brian Allison, Associate Designer
Photography by Geza Szurovy
Cover design: Theresa A. Twigg
Cover photograph: Brent Blair, Harrisburg, Pa.

WK2
0629935

Contents

Introduction

The most effective aviation savings tip of all time is: Don't fly! You will save mountains of money, dramatically increase quality family time, acquire new levels of respectability in the community for your sound fiscal judgment . . . and write off the most exciting time of your life. But if you do have enough right stuff to discard this savings idea outright, in spite of shortages in the green stuff, read on. There just might be a way to afford flying and send the kids to camp.

The cost of flying is growing by leaps and bounds. You could buy a brand-new four-seat fixed-gear 180-hp touring airplane in 1978 for $30,000 and you could choose from almost a dozen manufacturers. Fifteen years later the cost of a comparable new airplane from the remaining few active manufacturers had increased dramatically: $140,000 for the only U.S. model, more for the foreign models. A brand-new two-seat basic trainer in 1978 cost $20,000. Foreign manufacturers making comparable trainers were selling the two-seaters 15 years later for the ridiculous price of $100,000–120,000.

Nor are cost increases restricted to new aircraft alone; used aircraft prices have also kept steady pace. The cost of pilot certificates and ratings, aircraft rentals, insurance, parts and accessories, maintenance, aviation gasoline, and other flying expenses are all

heading for the stratosphere, significantly outpacing any growth in the disposable income of most pilots.

But there is good news, if you are determined to keep on flying or to learn and if you are willing to find every way you can to get the most for your money. During the generous years, we have built so much fat into our flying habits that there is tremendous scope to get more, a lot more, for your flying dollars, if you know where to look.

Cutting the Cost of Flying shows you how to slash thousands of dollars from your aviation budget and still do the kind of flying you covet so much. This book is for everyone who wants to make every penny count:

- Experienced pilots who own their own aircraft and are feeling the pinch of mounting bills.
- Student pilots being priced out of flying by the high cost of learning.
- Renter pilots concerned about the relentless increase of hourly rates.
- Pilots with freshly minted certificates, wondering how they can afford to continue flying.

Cutting the Cost of Flying presents more than 200 specific, money saving ideas, covering every aspect of private flying, gathered with the sole aim of keeping you in the air. Among other topics, we will cover:

- How to accurately measure all the costs of flying, to clearly understand where the savings opportunities are.
- How to buy an airplane with the performance and specifications you seek, for thousands of dollars below the price most pilots pay.
- How to learn to fly for half the cost paid by most student pilots.
- How aircraft partnerships slash the cost of flying.
- How to get others to legally pay for your private flying without violating "flying for hire" rules.

- How and where to get budget rental rates.
- How to economize on insurance costs without compromising coverage.
- How to pay less for tiedown and hangar space.
- How to get aviation supplies, parts, and accessories, and more at wholesale prices: engine oil, vacuum pumps, flying apparel, aviator sunglasses.
- How to slash the cost of maintenance without compromising flying safety, by finding a good mechanic appreciative of your budget objectives.
- How to save more maintenance dollars as an aircraft owner by doing your own preventative maintenance as permitted by FAR Part 43.
- How to make effective homemade alternatives for overpriced aviation accessories, such as wheel chocks, window shields, and IFR hoods.
- How to realize big savings by operating the aircraft at maximum efficiency.
- How to wring dollars out of the FARs.

Each savings idea is presented in two parts: an "action" section describes what the idea is, why it saves money, and how to implement the idea; a "savings" section quantifies in dollars the range of typical savings the idea can generate.

Savings is measured against the cost of other options, and has to be looked at in comparison to the alternative you are relying on now. Sometimes there is a range of differently priced options, and the option best suited to meet your needs might not be the least expensive choice. But if it is less expensive than your present arrangement, it represents savings from your perspective.

Some savings are only a few dollars; others are literally in the thousands. Some are clearly identifiable one-time savings. Others depend on how much you fly, what equipment you fly, and where you fly. Most are direct savings based on specific changes to your present flying practices and arrangements, but some, such as activist support for aviation interests, result in indirect returns over time, and benefit the entire aviation community.

The ideas for saving money are grouped according to typical aviation activities: buying an aircraft; renting an aircraft; buying supplies, equipment, and accessories; aircraft operations; aircraft maintenance; major overhaul and repairs; learning to fly; taking advantage of the FARs; and ideas that don't readily fit in other categories.

A few ideas that have several different applications are presented from different perspectives in several sections to make explicitly clear all the savings opportunities they offer.

Where appropriate, "**A word of warning**" or "**Warning**" alerts you to the potential pitfalls of an idea from the standpoint of safety, regulations, or other concerns. A few general words of warning are also in order:

- ➢ Always be extremely careful never to compromise flying safety in any way when you implement any of these ideas. When in doubt, seek professional advice.
- ➢ Take great care never to break Federal Aviation Regulations as you apply these ideas to your specific situation.
- ➢ Be very careful not to bite off more than you can chew regarding owner-performed preventative maintenance and the construction of homemade alternatives to commercially offered products. It can be very expensive to find out too late that you have exceeded your "do it yourself" limits.
- ➢ For best maintenance results, never be shy to get several quotes on any maintenance work to be done. Let it be known beyond any doubt that you are on a tight budget.

Clearly, not every idea will apply to everyone, but practically everyone will find enough new ideas to save hundreds and even thousands of dollars. Let's look at a typical example to see the potential for savings.

We will compare the flying costs of two pilots, one who spends freely but not foolishly, and another who does everything to cut flying expenses to the bone. The scenario is this:

- Each pilot learns to fly.
- After obtaining the pilot certificate, each pilot buys a four-seat, 180-hp, fixed-gear, fixed-pitch propeller touring airplane.
- Each pilot flies 150 hours during the first year of owning the airplane.
- Each pilot buys a reasonable amount of flying supplies and accessories during the first year of aircraft ownership.

Let's call the well-heeled pilot Jake, and the budget flier Chris.

Learning to fly

Take a look at Table 1. Jake got his license at a major FBO. Chris trained at a nonprofit flying club. Chris's certificate expenses of $2,890 represent a saving of $5,445 over Jake's expenses of $8,335, a surprisingly large amount. Two factors account for the bulk of the savings:

- Jake paid $70 an hour for the trainer. Chris paid only $35. This is not only because Chris flew at the nonprofit club that has lower rental rates, but also because Chris was content to train in the club's least expensive two-seat trainer, while Jake opted for the FBO's more modern four-seat training/touring aircraft. Chris also got a $5 per hour break on the instructor's hourly rate compared to Jake.
- An equally important factor was that Chris kept at it consistently, obtaining the pilot's certificate in 50 hours, while Jake, who didn't make the time to fly several times a week, needed 80 hours.

These two factors together accounted for $4,850 of Chris's savings on obtaining the certificate, compared to Jake's costs. Other significant savings came from being more frugal with ground school and learning aids.

Table 1

Learning to Fly

Major FBO		Nonprofit flying club		Savings
Certificate obtained in 80 hours due to commitments interrupting regular flying.		Certificate obtained in 50 hours as a result of flying regularly.		
Modern four-seater at **$70** per hour	$5,600	Old but sound two-seat trainer at **$35** per hour	$1,750	$3,850
Instructor at $25 per hour	$2,000	Instructor at $20 per hour	$1,000	$1,000
Didn't choose to get block time discount	$0	Less 10% block time discount	–$175	$175
Individual ground school, 15 hours at $15 per hour	$225	Group ground school for a flat course fee	$125	$100
Bought videos	$150	Borrowed videos	$0	$150
Bought computer ground school package	$140	Ground school materials included in course	$0	$140
Examiner's fee and aircraft rental for check ride	$220	Examiner's fee and aircraft rental for check ride	$190	$30
Total cost	**$8,335**	**Total cost**	**$2,890**	**$5,445**

The nonprofit flying club in this example did not charge an entry fee and monthly dues. Most clubs do, but the charges rarely exceed $500–$750 annually, still resulting in substantial savings over the FBO option.

Buying the aircraft

Upon completion of their certificates, Jake and Chris bought aircraft to maximize their freedom to fly when and where they pleased. Both pilots had exactly the same performance and specifications requirements, as noted in Table 2, and both pilots ended up with aircraft that could carry the same number of people at the same speed for the same distance. Both pilots financed 50 percent of the

purchase cost. But Chris spent only $48,210 (including the finance charges to be paid during the life of the loan), which is a whopping $76,928 less than what Jake spent. Let's see the sources of savings:

Table 2

Aircraft Purchase*

Five-year-old airplane		**Twenty-year-old airplane**		**Savings**
Fixed gear, fixed pitch 180-hp single-engine four-seater, zero-time engine, cruise speed 120 kts		Fixed gear, fixed pitch 180-hp single-engine four-seater, zero-time engine, cruise speed 120 kts		
Purchase price, no bargaining	$107,200	Purchase price, heavy bargaining	$43,500	$63,700
Financed 50% via a five year aircraft loan at 12% for total interest costs of:	$17,938	Financed 50% via a five year home equity loan at 8% for total interest costs of:	$4,710	$13,228
Total cost	**$125,138**	**Total cost**	**48,210**	**$76,928**

*Aircraft have same specifications and performance.

- Jake bought a five-year-old low-wing airplane with zero-time engine. Chris bought a twenty-year-old high-wing airplane with zero-time engine, new accessories, and new paint and interior. Both aircraft are powered by exactly the same type of engine, obtained from the same overhaul facility. Both are in excellent mechanical and cosmetic condition. Yet Chris's airplane cost $43,500 while Jake's cost $107,200. Yes, Jake's airplane is slightly more plush and modern in cosmetic appearance, but certainly not $63,700 more. Jake, perhaps influenced by how quickly cars become obsolete in contrast to airplanes, overlooked the fact that the only thing really old about Chris's airplane is the airframe structure, and overpaid considerably for the performance he bought.
- Chris also realized substantial savings on lower finance charges on the much smaller loan amount. Additional savings come from the lower interest rate of Chris's home equity loan (in addition, the interest on the home equity loan is deductible from income at tax time) compared to Jake's direct aircraft

loan. Do note what a huge chunk of money finance charges represent for both Chris and Jake—a good argument for minimizing borrowings.

Note, also, that while over the life of the loan the finance costs have to be paid, these costs are not cash out of pocket up front. The up-front cash that both pilots had to carry on the day of purchase was 50 percent of the purchase price, $53,600 in Jake's case, $21,750 for Chris. The bank provided the rest to the seller. Jake and Chris will be making repayments to the bank over the five-year life span of their loans.

High as the difference is in price between the two ownership alternatives, it is a much more common scenario than you might think. Check out the used aircraft prices if you need convincing. Jake and Chris could have further reduced the costs of acquiring their respective airplane significantly by buying it with one or more partners.

The first year of flying

Jake and Chris flew each airplane 150 hours during the first year of ownership. Table 3 is a detailed breakdown of their fixed and operating costs. By the end of the year, Chris had shelled out $15,020 less than Jake. Let's see why:

Table 3

The First Year of Ownership

Fixed costs		**Fixed costs**		**Savings**
Hangared at a regional airport $150 per month	$1,800	Tied down for $35 per month at a small field, a 40 minute drive from the regional airport	$420	$1,380
Insurance	$2,700	Insurance	$1,200	$1,500
State fees	$150	State fees	$150	$0
Loan payments of $1,192 per month on the 5 year, $53,600 aircraft loan at 12%	$14,304	Loan payments of $441 per month on the 5 year, $21,750 home equity loan at 8%	$5,292	$9,012

Table 3 **Continued**

Fixed costs		**Fixed costs**		**Savings**
Annual inspection fee, big FBO	$850	Annual inspection fee, independent mechanic	$450	$400
Total fixed costs	**$19,804**	**Total fixed costs**	**$6,432**	**$13,372**
Operating costs		**Operating costs**		
The aircraft flies 150 hours per year		*The aircraft flies 150 hours per year*		
Fuel bought at an average price of $2.00 per gallon at expensive large airports (fuel consumption, 8 gallons per hour)	$2,400	Fuel bought at an average price of $1.60 per gallon at small airports (fuel consumption, 8 gallons per hour)	$1,920	$480
Oil at an average price of $3.50 per quart bought at expensive FBO's can by can (consumption is 1 quart per 8 hours)	$66	Oil at an average price of $1.50 per quart bought at bulk from wholesaler (consumption is 1 quart per 8 hours)	$28	$38
Engine overhaul reserve, $7.50 per hour	$1,125	Engine overhaul reserve, $7.50 per hour	$1,125	$0
Maintenance, including three oil changes, new brake pads and a new vacuum pump, performed at an expensive FBO	$1,850	Maintenance, including three oil changes, new brake pads and a new vacuum pump, performed by an independent mechanic with owner participation	$720	$1,130
Total operating costs	**$5,441**	**Total operating costs**	**$3,793**	**$1,648**
Total year 1 costs	**$25,245**	**Total year 1 costs**	**$10,225**	**$15,020**

➢ It is after you sign the aircraft loan on the dotted line that the immense expense of borrowing becomes apparent. Jake's loan is much bigger than Chris's, and so are his monthly payments: $1,192 compared to $441. Chris's annual savings on loan repayments and finance charges in comparison to Jake's expenses is $9,012. Loan payments and finance charges also account for approximately half of each pilot's annual flying expenses.

- Over and above the expense of borrowing, Chris still comes out about $6,000 ahead, mostly on fixed expense savings. Lower hull insurance costs on the less expensive airplane and low tiedown fees at the small local field compared to Jake's expensive hangar at the big, regional airport account for $2,880. Another $400 comes from a less expensive annual inspection fee conducted by a mechanic specializing in working with owners to control maintenance budgets.
- Operating savings come from several sources: less expensive avgas (there are big variations in avgas prices at the pump), engine oil bought in bulk, and lower maintenance expenses. The maintenance savings are realized two ways: working with a mechanic who assists budget owners and owner-performed preventative maintenance allowed under FAR Part 43.

So there it is: the same amount of flying hours at the same speed transporting the same number of seats over the same distance. One pilot paid $25,245 in annual expenses; the other pilot paid only $10,225. Even if you fall somewhere in between Jake and Chris, it is easy to see how you, too, can wring at least a couple of thousand dollars in savings out of your annual flying costs without suffering the least bit in flying quality.

Buying supplies and accessories

Jake and Chris acquired a variety of supplies and accessories that are not crucial to flying but are nice to have. Table 4 shows the results buying niceties for their first year of ownership. By shopping carefully, ferreting out discount sources, and opting for homemade solutions, Chris spent $640 less.

Miscellaneous First Year Aviation Expenses Table 4

Expensive source		Budget source		Savings
Pilot's kneeboard	$29.95	Clipboard from discount store	$3.95	$26.00
Window heat shields	$75.00	Homemade heat shields	$30.00	$45.00
Aviator sun glasses	$135.00	Aviator sun glasses	$74.00	$61.00

Table 4

Continued

Expensive source		Budget source		Savings
Wheel chocks	$17.50	Homemade wheel chocks	$3.00	$14.50
Leather flying jacket	$300.00	Leather flying jacket	$170.00	$130.00
Leather flight bag	$220.00	Leather flight bag	$95.00	$125.00
Gust locks	$13.50	Homemade gust locks	$4.95	$8.55
Relief bottle	$4.50	Surplus jar	$0	$4.50
Fuel sampler	$6.95	Baby food jar	$0	$6.95
Maglite flashlight	$15.95	Maglite flashlight	$9.95	$6.00
Headset	$210.00	Earphone	$6.95	$203.05
Pitot tube cover	$9.95	Homemade pitot tube cover	$0	$9.95
Total Expenses	**$1,038.30**	**Total Expenses**	**$397.80**	**$640.50**

Overview

Table 5 summarizes how the pilots fared from being student pilots through their first year of aircraft ownership. Both pilots bought equally capable airplanes and both had exactly the same flying experience. Yet one spent $88,218 while the other spent only $35,263. This example is not extreme. The difference of $52,955 dramatically illustrates the extent to which flying expenses can be controlled with a little effort and ingenuity.

Table 5

Summary Cash Savings—Pilot's Certificate and first Year of Aircraft Ownership

Cash cost of pilot's certificate and first year of aircraft ownership*	Expensive option	Budget option	Savings
Private pilot's certificate	$8,335	$2,890	$5,445
Up front cash cost of purchase of 180 hp fixed gear, fixed pitch four seat aircraft (50% of purchase price, which is	$53,600	$21,750	$31,850

Continued Table 5

financed, and finance charges, neither due up front, are excluded).			
First year fixed and operating expenses (includes first year loan payments; aircraft fly 150 hours during the year)	$25,245	$10,225	$15,020
First year miscellaneous aviation expenses	$1,038	$398	$640
Total expenses and savings	**$88,218**	**$35,263**	**$52,955**

*Expensive and budget aircraft have the same specs and performance.

There are endless variations on the ways to cut flying costs. Expenses could be further reduced through a partnership, or by flying a less-expensive two-seater, or by flying fewer than the 150 hours in the first year of ownership—but under all circumstances continuing to fly. So read through this book and see what works best for you to get to the expense levels you can afford. Use a good dose of common sense, be creative, and enjoy it up there.

An early ambitious attempt at budget transportation that didn't go too far.

1

Buying an aircraft

1 Borrowing—how much to borrow and for how long?

✻ **Action** Do not borrow any money to buy an airplane, or, keep the amount you borrow and the time that you borrow it to absolute minimums. Borrowing is renting money. The interest payment is the rental fee. The percentage of interest charged for a loan might seem a small amount, but the total interest you will have to pay over the life of the loan can be a staggering amount, much more than most people suspect when they are initially informed of the modest sounding annual percentage rate.

In addition to being expensive, the typical loan repayment schedule is structured to repay most of the interest up front to give the lender its income from the loan as soon as possible. Of each fixed monthly repayment, initially most of the amount is applied to reduce total interest owed and very little to reduce principal owed. As the loan is paid down, the amount applied to principal increases only gradually.

As a result, for a disproportionately long time, large loan principal amounts remain outstanding even though monthly payments are

being made. In case you want to prepay, say, halfway through the loan's life, you will find that you still owe a surprisingly large principal amount because most of your payments were applied to interest.

Ask your lender for a printout of a full repayment schedule disclosing amounts of each payment applied to interest and principal for each loan amount and maturity you are considering. Table 1-1 is an example.

Table 1-1 **Amortization table**

Loan amount	$ 25,000.00
Interest rate	12.00%
Loan term years	10
and months	0
Monthly Payment	$ 358.68

Month of loan	Months interest	Months principal	Remaining balance
1	250.00	108.68	24,891.32
2	248.91	109.77	24,781.55
3	247.82	110.86	24,670.69
4	246.71	111.97	24,558.72
5	245.59	113.09	24,445.63
6	244.46	114.22	24,331.41
7	243.31	115.37	24,216.04
8	242.16	116.52	24,099.52
9	241.00	117.68	23,981.84
10	239.82	118.86	23,862.98
11	238.63	120.05	23,742.93
12	237.43	121.25	23,621.68
13	236.22	122.46	23,499.22
14	234.99	123.69	23,375.53
15	233.76	124.92	23,250.61
16	232.51	126.17	23,124.44
17	231.24	127.44	22,997.00
18	229.97	128.71	22,868.29
19	228.68	130.00	22,738.29
20	227.38	131.30	22,606.99
21	226.07	132.61	22,474.38
22	224.74	133.94	22,340.44
23	223.40	135.28	22,205.16
24	222.05	136.63	22,068.53
25	220.69	137.99	21,930.54
26	219.31	139.37	21,791.17
27	217.91	140.77	21,650.40
28	216.50	142.18	21,508.22
29	215.08	143.60	21,364.62

Month of loan	Months interest	Months principal	Remaining balance
30	213.65	145.03	21,219.59
31	212.20	146.48	21,073.11
32	210.73	147.95	20,925.16
33	209.25	149.43	20,775.73
34	207.76	150.92	20,624.81
35	206.25	152.43	20,472.38
36	204.72	153.96	20,318.42
37	203.18	155.50	20,162.92
38	201.63	157.05	20,005.87
39	200.06	158.62	19,847.25
40	198.47	160.21	19,687.04
41	196.87	161.81	19,525.23
42	195.25	163.43	19,361.80
43	193.62	165.06	19,196.74
44	191.97	166.71	19,030.03
45	190.30	168.38	18,861.65
46	188.62	170.06	18,691.59
47	186.92	171.76	18,519.83
48	185.20	173.48	18,346.35
49	183.46	175.22	18,171.13
50	181.71	176.97	17,994.16
51	179.94	178.74	17,815.42
52	178.15	180.53	17,634.89
53	176.35	182.33	17,452.56
54	174.53	184.15	17,268.41
55	172.68	186.00	17,082.41
56	170.82	187.86	16,894.55
57	168.95	189.73	16,704.82
58	167.05	191.63	16,513.19
59	165.13	193.55	16,319.64
60	163.20	195.48	16,124.16
61	161.24	197.44	15,926.72
62	159.27	199.41	15,727.31
63	157.27	201.41	15,525.90
64	155.26	203.42	15,322.48
65	153.22	205.46	15,117.02
66	151.17	207.51	14,909.51
67	149.10	209.58	14,699.93
68	147.00	211.68	14,488.25
69	144.88	213.80	14,274.45
70	142.74	215.94	14,058.51
71	140.59	218.09	13,840.42
72	138.40	220.28	13,620.14
73	136.20	222.48	13,397.66
74	133.98	224.70	13,172.96
75	131.73	226.95	12,946.01
76	129.46	229.22	12,716.79
77	127.17	231.51	12,485.28

Table 1-1 **Amortization table continued**

Month of loan	Months interest	Months principal	Remaining balance
78	124.85	233.83	12,251.45
79	122.51	236.17	12,015.28
80	120.15	238.53	11,776.75
81	117.77	240.91	11,535.84
82	115.36	243.32	11,292.52
83	112.93	245.75	11,046.77
84	110.47	248.21	10,798.56
85	107.99	250.69	10,547.87
86	105.48	253.20	10,294.67
87	102.95	255.73	10,038.94
88	100.39	258.29	9,780.65
89	97.81	260.87	9,519.78
90	95.20	263.48	9,256.30
91	92.56	266.12	8,990.18
92	89.90	268.78	8,721.40
93	87.21	271.47	8,449.93
94	84.50	274.18	8,175.75
95	81.76	276.92	7,898.83
96	78.99	279.69	7,619.14
97	76.19	282.49	7,336.65
98	73.37	285.31	7,051.34
99	70.51	288.17	6,763.17
100	67.63	291.05	6,472.12
101	64.72	293.96	6,178.16
102	61.78	296.90	5,881.26
103	58.81	299.87	5,581.39
104	55.81	302.87	5,278.52
105	52.79	305.89	4,972.63
106	49.73	308.95	4,663.68
107	46.64	312.04	4,351.64
108	43.52	315.16	4,036.48
109	40.36	318.32	3,718.16
110	37.18	321.50	3,396.66
111	33.97	324.71	3,071.95
112	30.72	327.96	2,743.99
113	27.44	331.24	2,412.75
114	24.13	334.55	2,078.20
115	20.78	337.90	1,740.30
116	17.40	341.28	1,399.02
117	13.99	344.69	1,054.33
118	10.54	348.14	706.19
119	7.06	351.62	354.57
120	3.55	355.13	−0.56

Total interest paid	$ 18,041.02
Loan amount	$ 25,000.00
	$ 43,041.02

If you must borrow a specific amount to get you in the air, keep the maturity to an absolute minimum to keep total interest paid as low as possible.

$ **Savings** Hundreds or thousands of dollars in interest are saved. To get an idea of just how expensive borrowing is, consider a $17,500 loan (half the purchase price of a $35,000 airplane) for 10 years at an annual interest rate of 10 percent. Your monthly payments (principal and interest) on these terms are $231.26. This does not seem to be such a large amount, does it? Well, by the end of the 10 years, you will have made total repayments of $27,752 of which interest payments are $10,252. That is $10,252 straight out of your pocket never to be seen again, equal to almost a third of the value of the airplane when you bought it!

If you can manage to scrounge up another $10,000, limiting your borrowing to only $7,500 for 10 years at 10 percent, your total interest paid will be $4,394, an improvement but still a lot of money.

The interest expense also improves if you borrow the full $17,500 at 10 percent but can reduce the maturity to 5 years by making larger monthly payments. Under these terms your total monthly payments are $371.82 (compared to $231.26 with the 10-year maturity), total repayments over the life of the loan are $22,309, of which total interest payments are $4,809 (compared to $10,252 over 10 years).

2 Borrowing—the home equity option

✻ **Action** If you must borrow, do it through a home equity loan if you can, instead of a regular consumer loan. A home equity loan is a second mortgage on your house and it offers two savings possibilities. First, interest rates for mortgages, even second mortgages, are usually lower than rates for other comparable consumer loans. Second, the interest on home equity loans is tax deductible.

A nonfinancial advantage of a home equity loan is that it is secured by a lien on your house rather than a lien on the airplane; thus, if you have enough equity in your house and you so choose, you can finance 100 percent of the airplane's purchase price, something you couldn't do if you took out a loan secured by the airplane.

$ **Savings** Hundreds or thousands of dollars over a direct consumer loan are saved, depending on terms and amount.

As an example, let's say that you want to borrow $10,000 for 10 years. The consumer loan is at an annual interest rate of 10 percent, the home equity loan is at 8 percent. Your total interest expenses for the consumer loan will be $5,858; total interest expenses for the home equity loan will be $4,559, a savings of $1,299, prior to tax benefits.

Additional savings will result in the tax deductibility of the interest on the home equity loan, depending on the tax bracket you are in. In the example, your total interest expenses for the home equity loan is $4,559. You can reduce your taxable income over the life of the loan by this amount (deducting the actual amount of interest paid in each year from that year's income). Assuming that you are in the 28 percent tax bracket, over the life of the loan, your tax savings will be 28 percent of $4,559, or $1,276. In comparison to the straight consumer loan, on an after-tax basis, your total savings will be $2,575 ($1,299 + $1,276).

3 Borrowing—the private loan option

* **Action** Another alternative to reducing the cost of borrowing in comparison to a consumer loan is to arrange a loan from a private individual at a lower rate. Private individuals are willing to make these loans because they can earn more on lending the money to you (secured by the airplane) than by investing it in other low-risk opportunities such as bank deposits.

Don't expect just any private individual to lend you money. Whoever would be willing to do so will have to know you and your financial circumstances very well and understand the airplane business to feel comfortable with the underlying risk. Typical examples of the private loan option are loans from relatives, or from one partner to the other.

If you do choose this option and find a willing private lender, be sure to transact the loan entirely at arms length in a professional manner documented by a note and properly secured by a lien on the airplane. It will save you both a lot of headaches in case of any sort of misunderstanding or dispute down the road.

$ **Savings** Hundreds or thousands of dollars in lower interest saved. Savings will obviously depend on loan terms and amount. As an example, let's again assume that you want to borrow $10,000 for 10 years. The consumer loan is at an annual interest rate of 10 percent, the loan from the private source is at 8 percent. Your total interest expenses for the consumer loan will be $5,858; total interest expenses for the private loan will be $4,559, a savings of $1,299.

Private lenders find this transaction attractive because their gain in comparison to a commercially available alternative investment at 6 percent will be in the same ballpark as your savings in comparison to your alternative, so you both come out ahead.

Buying at auction

✻ **Action** Buy an aircraft at auction. Dozens of aircraft get auctioned off every year. The Drug Enforcement Agency, marshal's offices, and banks regularly auction confiscated or repossessed aircraft. Although getting a deal is not as easy and simple as the many auction ads might suggest, for some pilots an auction might present a golden budget opportunity.

Avoid publications—sold for a fee—that promise to lead you to auctions nationwide. By the time you would get to those opportunities, chances are professional dealers have picked out the best bargains. Stick to specific auctions advertised in publications such as *Trade-A-Plane*. Do send in for the information packages on individual items of interest to you that are listed in the announcement of a specific auction.

A word of warning: This is an option only for people who really know what they are doing when it comes to evaluating mechanical condition and a fair price. The main problem is that in most cases little is known about the history of the aircraft. Many of them don't even have logbooks. Practically all of them have not flown for ages before the auction. It can take a lot of time-consuming litigation and paperwork from confiscation or repossession to auction; months and even years.

Thus, most of the bidders at auctions are savvy wholesalers and repair shops who know exactly what needs to be done to make the aircraft being auctioned resalable, and who will not pay a penny more than what constitutes a real bargain. The trick for you, the retail purchaser, is to hook up with a professional mechanic you can trust, who has experience buying at auctions, knows what you want, and can tell you what it will cost including restoration to get an offered aircraft back into documented mainstream use.

If you know what you are doing, you can get an excellent deal if on the day of the auction there happens to be little demand for the aircraft you want. The auctioneer might set a minimum bid below which the aircraft will not be sold, but this amount generally depends on what is owed to secured creditors, and can be very low in comparison to the aircraft's true value.

One drawback of auctions is the expense of getting to them if you are not in the immediate area. There is no assurance that you will get what you are looking for right away, and travel expenses can mount quickly.

$ **Savings** Thousands of dollars can be saved over the price of comparable aircraft bought through alternative means.

5 Buying damaged aircraft

✳ **Action** Buy a damaged aircraft at a deep discount, and have it repaired. For some pilots, damaged aircraft can represent a savings opportunity if the seller is anxious to get rid of the wreck and is

willing to underprice it; however, it cannot be overemphasized that you really have to know what you are doing.

You have to be able to know exactly what the extent of the damage is, what it will cost to repair to a condition as good as new, and what effect the damage history will have on resale value when you want to sell the airplane. Only then can you figure out the price to pay to get a real bargain.

It is essential to have an excellent relationship with a trusted mechanic who is qualified to assess the damage and do a top quality repair job for the quoted price. If your choice of mechanic backfires, it can be a costly mistake.

$ **Savings** Thousands of dollars can be saved if the acquisition price and the mechanic's repair assessment are right.

6 Certified aircraft vs. experimental aircraft

* **Action** Consider purchasing an experimental aircraft instead of a production aircraft. Evaluate the performance of production and experimental aircraft in the same category and research the respective prices. If the price for aircraft of comparable performance is the same or the experimental aircraft costs less, the experimental aircraft might be the better buy. The chief advantage of the experimental aircraft is that you can do much more work on it yourself; thus, this option is best suited for pilots with a mechanical bent, skilled in working on aircraft.

Building standards in the homebuilt/kitbuilt community are high and the safety record is good; however, there is some inconsistency in construction quality. The difficulty in buying an experimental aircraft is accurately evaluating what you are getting. A good mechanic who is knowledgeable about homebuilts is indispensable. The Experimental Aircraft Association has excellent guidelines for prepurchase inspections, and a corps of technically qualified volunteers to guide you. Specific clubs for a particular type also provide strong support. (*See* Resources.)

$ **Savings** Hundreds of dollars or more can be saved on the purchase price and on subsequent maintenance savings.

7 Direct purchase vs. aircraft brokers

✳ **Action** Purchase your used aircraft directly from its owner instead of from an aircraft broker. At a minimum you will save the broker commission (usually 5 percent). Most likely you will save more because, like anyone in professional sales, brokers are quite aggressive in maximizing the sales price of aircraft.

Having in-depth industry knowledge and sales experience, brokers are adept at unleashing a barrage of convincing sounding arguments that often succeed in pushing the unwary buyer into a transaction at the high end of the price range. Brokers are also usually more successful in conveying a "take it or leave it" attitude than an anxious owner, especially if they sense the slightest nibble.

The vast majority of brokers are fundamentally honest business people and provide an important service that for many buyers and sellers is the best way to go. A broker is the appropriate choice for sellers who do not know how to sell their airplane or don't have the time to do so. Often a seller will not be out of pocket having to pay the broker's commission because the broker will be able to sell the airplane for considerably more than the seller could have done.

Brokers are also a good choice for buyers who want an airplane now with the least hassle and are willing to pay a premium. Expensive corporate aircraft are best sold through reputable brokers. A specialist such as a warbird broker can be of great service in ferreting out a rare find. But for the well-informed, budget-conscious, light aircraft buyer who is willing to search diligently for the right airplane and drive a hard bargain, a broker in the middle will usually mean an extra expense.

Don't take these pages at face value. See for yourself. When you conduct your aircraft search, carefully document each airplane that appeals to you and compare the prices of similar aircraft offered directly by owners and by brokers. You will find plenty of instances where the broker prices are considerably higher.

$ **Savings** If you are well-informed and know exactly what the aircraft you are buying is worth, you should be able to save about 5 percent of the purchase price by buying directly from an owner. This is not a small sum even in the case of the least expensive aircraft. The broker's commission on a $12,000 Piper Cub is $600; a $50,000 Cessna Skylane commission is $2,500. In many instances you might be able to save a lot more by bargaining with a seller who isn't relying on a broker to tell her how much she should be getting for her airplane.

8 Fixed-gear vs. retractable-gear

✻ **Action** Buy a fixed-gear airplane of comparable performance instead of a retractable-gear airplane. You can realize great savings in maintenance costs as well as in the generally lower purchase price of the equally able but less sexy fixed-gear machine.

In the case of light aircraft, it is generally accepted that the additional weight of the retractable gear mechanism negates the positive effect of reduced drag. The net gain in performance is thus marginal at best, and meaningless on the shorter trip lengths flown by most light aircraft.

$ **Savings** Savings on a fixed-gear aircraft over a comparable retractable can be several thousand dollars. In addition to savings on purchase price, annual savings of between $50–$150 can be easily realized in lower inspection fees because the fixed-gear aircraft does not need the rather involved retraction test. The fixed-gear option will save hundreds of more dollars if the retractable-gear option needs actual maintenance over and above the periodic inspections and retraction tests. As an example, the universal joints in the Piper PA28 and PA32 series gear-retract mechanisms, which eventually might need replacing, cost more than $500 in parts alone.

If you are unfortunate enough to collapse a nose gear—not an unusual event in some retractable gear aircraft types known for a notorious weakness in the nose gear mechanism—or make a gear-up landing, your savings in the fixed-gear alternative could be significant.

9 Hangaring—buying a hangar

✲ **Action** Buy a hangar condominium if you can afford it. The price range for a T-hangar condo might be around $20,000. How, you might ask, do I save money by shelling out $20,000? You save because when you no longer need the hangar condo and you sell it, you essentially get back all, more, or most of the money you would have forked over in hangar fees during the period you owned the hangar condo.

Most likely, as general aviation airports and airport space become increasingly scarce, you would be able to sell the hangar for quite a bit more than it cost you to buy.

$ **Savings** Potentially thousands of dollars could be saved. Suppose that hangar space at your airport rents for $120 per month. You buy a hangar for $20,000 and sell it in five years for $25,000. You had free hangaring for five years that otherwise would have cost you $7,200. In addition, your investment appreciated by $5,000. If you finance the hangar, you also have to factor in expensive finance charges that might only allow you to break even while the loan is outstanding; however, when it is paid off, you are ahead.

10 Hangaring—location

✲ **Action** If you are set on hangaring your airplane, find the airport with the least expensive hangar space in your area even if it means driving a fair distance. You are in the budget game so you should be willing to drive a distance that does not burn up the hangar savings.

$ **Savings** Typically $30–$100 per month, or, $360–$1,200 per year could be saved. Rates vary to the greatest extent around large population centers. You will be surprised to find how quickly the rates drop as soon as you get as little as 10–20 miles away from the big general aviation airports in and around large cities.

11 Hangaring—share a hangar

✻ **Action** An excellent way to cut down on hangar costs is to find compatible aircraft and owner(s) to share a T-hangar. It is not uncommon to see two to three aircraft sharing a T-hangar if the space is large enough. Good combinations are a production light single and a smaller airplane such as a Pitts Special or a Vari-Eze. A hangar can also be shared with nonaviation vehicles. Boaters and antique car owners seeking storage for the winter season are always good candidates to spread around the rent a bit. (Fig. 1-1)

Figure 1-1

Three aircraft share this hangar. Note the sailplane suspended from the roof by an electric winch, a common arrangement in some of the more expensive parts of the world.

Always check the lease or rental agreement to determine occupancy restrictions. Not all airports allow the sharing of T-hangars and some airports might charge higher shared rates. Check it all out carefully before committing to anything.

$ **Savings** Perhaps you will save \$40–\$160 per month per person, or \$480–\$1,920 per year, depending on the total cost of the hangar and the number of people sharing. Three people sharing a \$250 per month hangar will each pay \$83 per person, a savings of \$167 per person over the single-renter rate. Through hangar space sharing, hangar costs per person can be reduced to levels where they become competitive with tiedown costs.

Hangaring vs. tiedown

✻ **Action** An argument can be made that even if you have the money to hangar your airplane, tie it down instead anyway. Here is how it goes: A hangar is so much more expensive than a tiedown that if you tie down the airplane and put the monthly difference between the hangar and the tiedown rate in the bank, you can repaint your airplane approximately every three years (that is how the numbers seem to work out at many airports for a 4–6 seat single). Because most well-kept—albeit tied down—production airplanes will easily look practically new even with a five-year paint job, you can repaint every five years, end up with a brand-new looking airplane, and pocket two years worth of difference between the tiedown rate and the hangar rental rate.

$ **Savings** Typically you could save in the $700–$3,500 range assuming the airplane is repainted every three to five years. Say the monthly tiedown fee is $75 and the hangar fee is $180, a difference of $105 per month, or $1,260 per year. If in three years the tied down airplane is repainted for $3,000 (the cost of repainting a typical four-seater), the savings is $780 ($1,260 × 3–$3,000). If the tied down airplane is repainted after five years, the savings is $3,300 ($1,260 × 5–$3,000). In all cases, the tied down airplane freshly repainted at the end of the period will look better and will be worth more than the hangared nonrepainted airplane at the end of the same period.

13 Insurance—amount

✻ **Action** Obtain only the amount of insurance appropriate to your circumstances and applicable to your aircraft. There are two main components to aircraft insurance policies: hull insurance protecting the value of the aircraft and liability insurance protecting you, the owner, from lawsuits arising out of damage done by your aircraft. While there is not too much room to maneuver in buying insurance coverage, some savings are possible by carefully tailoring your policy to meet your specific needs.

Regarding hull insurance: Do not over-insure your airplane. The temptation to do so exists. Some people like to be very conservative and it is also nice to think of being handed a generous check after a loss and buying something slightly better. Typical agreed values acceptable to insurers are ±10 percent of retail blue book value, giving you a 20 percent spread in the insured amount you can choose. Insurance in excess of such values might be available if justified to the insurance company, but costs more and can be counterproductive. The insurance company might find it less expensive to repair the overinsured aircraft that is severely damaged than to declare it a loss and shell out the total insured amount.

Select a realistic hull insurance replacement value for your aircraft "as is." Don't try to build in the paint job and the radios you would like it to have, or insure it for the condition that it was in years ago. Do account at annual insurance renewal for any appreciation that might have taken place (adjusted for present condition of the powerplant, interior, and exterior). Given how few new aircraft are being built, appreciation does occur, even accounting for wear and tear. So, make adjustments but don't over-insure.

Regarding liability insurance: Insurers offer minimum liability coverage, but for most people this is rarely enough; however, people may also over-insure at additional expense. How much liability insurance to get is a function of your personal circumstances. If you are unmarried without children and have no assets besides the airplane, a fairly low amount might be sufficient. If you have a family, house, and investments, liability insurance is not the place to save on flying costs.

$ **Savings** Typically, $50–$300 per year might be saved. You can save up to a few hundred dollars by avoiding any pressure to over-insure, depending on hull insurance premiums for your type of aircraft and your specific liability insurance needs; however, take great care to never under-insure because the downside, especially in case of liability insurance, could wipe you out financially.

14 Insurance—competing quotes

* **Action** Get competing quotes if you are dealing directly with a national insurance underwriter; if you are insuring through a broker, make sure the broker gets a wide range of quotes to get the best price. When your broker gives you the best quote, call other brokers to see if they can better it significantly. Be aware that once you ask one broker to obtain quotes on your specific airplane other brokers can only get indicative quotes, but cannot get a specific quote on your airplane until the original broker releases the airplane. This is an insurance industry practice made necessary by the way specific quotes are underwritten. What other brokers can tell you, however, is whether or not your quote is reasonable.

Insurance rates can fluctuate widely from year to year. One reason to check out the market at renewal time is because not all insurers will pass on to you in renewal quotes a significant decline in rates. They hope you will be happy to see that your rate didn't increase and you will quickly send them a check.

$ **Savings** Competing quotes can save hundreds of dollars per year. Quotes from good brokers will usually be in the same ballpark, though expect some to be more competitive than others. Do your homework and remember that your broker is only as good as the underwriters he contacts, so make sure he is doing his job. Don't be surprised to see occasional differences of several hundred dollars if you are dealing directly with underwriters. Closely scrutinize, however, the terms and conditions of each policy. The lower the price, the more the small print can take away from what the big print promises.

15 Insurance—deductible

* **Action** Choose higher deductibles to realize some savings on insurance premiums.

$ **Savings** Expect to save $50–$100 per year depending on insured amount. Though the option to save on annual premiums by choosing higher deductibles has narrowed in recent years, policies are still out there on which some savings are possible.

 Insurance—seasonal use

* **Action** If you do not fly your aircraft throughout the year, take it off in-flight insurance for the period during which you do not fly it. This arrangement can be especially attractive to owners of gliders, motorgliders, balloons, and other aircraft that are often flown only seasonally.

  Savings Hundreds of dollars could be saved. Insurance is priced proportionally over the insured period. With some minor exceptions, $1,000 worth of annual insurance costs $500 for six months. Your savings will be significant but somewhat less than that because you will want to keep not-in-flight insurance in effect throughout the year.

 Leasebacks

* **Action** Lease your airplane to a flight school or flying club to realize income on your airplane. Most privately owned aircraft are underutilized. Leasing them to other operators increases the airplane's utility and might produce some income for the owner, defraying the expenses of owning the aircraft.

  A word of caution about leasebacks. Most owners enter into leasebacks because they think they will make money on a net basis, a profit over and above what it costs them to own and operate the airplane. These expectations often turn out to be wildly optimistic and there is disappointment when the money received does not exceed the expenses of the airplane. The problem usually stems from the fact that the airplane flies less than what was estimated when the lease was set up.

  The real benefit of most leasebacks is not that they make a profit but that they make some significant amount of money for the owner, which substantially subsidizes the owner's expenses. To truly subsidize ownership of an airplane, a leaseback should at least pay more to the owner than the value of the wear and tear on the airplane per leased flying hour.

Leaseback terms have no hard and fast rules. Each leaseback agreement is entirely up to the lessee and lessor. Usually the owner receives a fixed hourly rate for the airplane and the two parties work out an agreement for such expenses as maintenance and parts. The operator leasing the airplane usually pays for fuel and tiedown costs. Agreement is also reached over the kind of access and any priority the owner will have to the airplane.

Leasebacks are not for everyone. Not all aircraft are good candidates for leaseback and for many owners an airplane is too personal a possession to be placed at the disposal of others. But owners of popular production airplanes who fly only a hundred hours or less annually and are looking for some way to recoup some of the expenses of ownership, a well-structured leaseback might be just the thing.

$ **Savings** Potentially hundreds or even thousands of dollars could be saved. Assume you receive free tiedown, free labor, and $25 per hour for leasing your airplane and it flies 150 hours a year under the leaseback program. The total amount you will receive in lease income for the year is $3,750. Of this amount, let's assume that wear and tear per hour is $15. (Carefully determine engine and maintenance reserve costs per hour for your airplane when setting the hourly lease rate.) The $15 is the value of the airplane used up during each leased flying hour, leaving $10 in your pocket; therefore, at 150 hours per year, you will have received net cash income of $1,500. In addition, free tiedown that would have otherwise cost $80 per month, and free labor amounting to a value in this example of $1,000, results in annual savings of $1,960. Together with the net cash income, the total annual value of the lease back is $3,460, a tidy subsidy toward your annual flying expenses.

18 Negotiating the purchase

⁂ **Action** You can realize considerable savings when buying an aircraft by making sure that you get a good deal. There is never one single right price for an aircraft. There are only guidelines and price ranges for aircraft of similar type, age, and with similar equipment. It is up to the buyer to negotiate a deal at the low end of this price range.

Extenuating circumstances such as a forced sale or ignorance on the part of the owner of the aircraft's true value might result in a sale priced significantly under market. Negotiating the best deal requires some work on your part. Be firm in bargaining, but be fair.

Always know the price range for the airplane you seek. This means doing your homework in great detail regarding condition, times on airframe, engine and accessories, airworthiness directive compliance, options, radios, other equipment, types of use (any training?), and damage history. A good source of current prices is the N.A.D.A. *Retail Aircraft Appraisal Guide* (which is available to everyone from N.A.D.A. Appraisal Guides P.O. Box 7800, Cosa Mesa, CA 92628, 800-966-6232) (Fig. 1-2).

Figure 1-2

The N.A.D.A. guide is one of several good sources of information on used aircraft prices.

Know what is hot and what is not, and develop your strategy accordingly. A hot-selling airplane type might not be such a bargain to begin with, so you might want to reconsider your choice of type.

But if a hot-seller is what you want, the ability to move quickly to check out the airplane and close the deal immediately at a fair price will do the trick. Be careful not to be swayed into overpaying as a result of competition for the airplane. If supply is greater than demand, you have a lot more time and room to bargain.

Never reveal up front the maximum amount you are willing to pay for an airplane. Always offer an initial amount that is a reasonable percentage less than the upper limit you set for yourself (a good rule of thumb is 20 percent). Develop specific well-researched and reasoned arguments and apply them aggressively to tell a seller why the asking price is unreasonable.

Have the cash available and any financing in place to close the deal immediately if the airplane checks out. Use this as an incentive to drive a hard bargain. A seller who "sees" the money is often motivated to give in to an extra discount to get the sale done.

Very carefully do a full mechanical inspection (performed by your mechanic) and note any plausible squawks, including all squawks not critical to flying safety. Have a generous repair estimate done for each item and aggressively demand that the costs be deducted from the selling price. Leave room for some compromise and you will invariably get some reduction in price. If the airplane's annual is due soon, use that as a bargaining chip.

Be aggressive in seeking reductions due to any damage history (carefully check out the adequacy of any repairs). While a truly professional repair job makes an airplane as good as new, the price is always depressed by damage history due to the perception that you can never be entirely sure of the adequacy of repair. Some owners are more willing to give in to such perceptions than others. If you know what you are really getting, but the seller doesn't, you can get quite a bargain.

Some sellers will offer to fix a squawk prior to sale instead of reducing the price by the cost of the fix. Do not accept the offer if the problem is critical to flight safety; let your mechanic do the repair and use her repair estimate to reduce the seller's asking price.

Look for signs of anxiety to sell. Sad as they are, a whole host of financial calamities suffered by the seller might result in a bargain basement price for the buyer.

If you reached a price below which the seller is unwilling to go, there is one more tactic you can try. Ask the seller to include at no additional cost any accessories that were not going to be sold with the airplane: a preheater, a portable loran or GPS receiver, a hand-held radio, a portable oxygen system, life vest, foldable bikes, and the like.

If you need financing, ask the seller for financing at a rate below the bank's loan rate but above the rate the seller could get on an alternative investment. You both win.

Don't waste time on blockheads. Some people ask an unreasonably stiff price and refuse to budge. Pass them by quickly, but contact them in a few months if you are still in the market. The longer an overpriced airplane is unsold, the greater the chance that the owner will see reason.

$ **Savings** Bargaining can save hundreds to thousands of dollars. The ability to negotiate a good deal will literally leave that much more cash available to put in your pocket or to put into improvements of the airplane after you have bought it.

Negotiating the sale—know the tricks in reverse

✻ **Action** When it comes time to sell the airplane, apply in reverse some of the negotiating skills you used to purchase it. Set a higher asking price than what you are willing to accept. Research the going price of comparable aircraft and be aggressive but not outrageous in setting your price. Establish in your own mind a selling price below which you won't go, and hold to it firmly. Avoid looking desperate to sell. Be cautious with financing the buyer; in principle don't do it, but if you have your reasons to provide financing, document it properly and take a lien on the airplane.

In any trade transaction, one side makes more money if the other side doesn't know the value of the product being traded. The key to selling your airplane for a high price is being able to wait for the uninformed, well-to-do buyer to show up.

$ **Savings** You could potentially save hundreds of dollars or more. You will maximize your sales price by your ability to negotiate the sale effectively and your ability to wait for the right buyer to turn up.

20 Partnerships

✻ **Action** Buy an aircraft in partnership with one or more co-owners. Your savings will be immense in comparison to owning alone. The savings are so significant that for many of us a partnership is the only way in which we can afford to own an airplane at all. The savings in partnerships are realized in two areas: aircraft acquisition cost and fixed costs.

Regarding aircraft acquisition cost, the amount of money you need to buy an airplane is greatly reduced—at least halved if you opt for only one partner. Four partners would each have to raise only one-fourth of the money required for a solitary buyer. Any financing needs per partner are also significantly less.

Regarding fixed costs, significant savings are realized on the equal sharing of the expenses that have to be paid regardless of whether or not the airplane flies: tiedown or hangar, insurance, annual, state fees, and loan payments. Fixed costs would probably be covered by a monthly dues structure that equally applies to each partner.

Note that per-person savings are not realized on hourly operating costs such as the fuel or oil and the hourly cash reserve for maintenance of the engine and airframe. These expenses are usually the sole responsibility of each partner during the hours flown by the partner. The hourly operating costs would be covered by each partner paying an hourly rate for time flown in a month; this is in addition to the regular monthly dues.

While partnerships offer significant savings, a partnership does not always mean that it is a financially better option than aircraft rental. That depends on the number of hours flown. The analysis presented in Tip 23, rental vs. ownership in this chapter still has to be performed. As a less expensive form of ownership, a partnership will reduce the minimum number of hours you have to fly per year to make ownership a financially attractive alternative to rental.

$ **Savings** A partnership can save thousands of dollars. Table 1-2 is a dramatic illustration of just how much you can save in a partnership in comparison to sole ownership. It shows the flying expenses of a typical used four-seat single for a sole owner and up to four partners. Pay particular attention to the aircraft purchase price per partner, the annual fixed expenses per partner, and the total hourly expenses per partner. This last category is the combined fixed and operating cost per hour per partner for 50, 100, and 150 hours flown per year.

As you can see, in a three-person partnership it is enough for a partner to fly only 50 hours per year (which results in a cost of $90 per hour) to be competitive with the hourly commercial rental rate of an equivalent airplane ($90) and to be way ahead of the comparative total hourly cost of sole ownership (a whopping $216).

Regarding the financial table and assumptions, the financial information presented is a generic example intended only to show you how to look at the depicted costs. Estimate your own costs based on information applicable to your own particular set of circumstances.

Note that the loan payments are the annual payments on the $22,500 loan calculated from the loan terms in the assumptions using your friendly banker's amortization formula (the formula is not shown).

In this scenario an annual fixed maintenance expense of $1,000 per year is set aside. This amount is over and above the hourly engine reserve and hourly general maintenance reserve, and provides an additional cushion. Some people choose to reserve for maintenance only in the per hour reserves.

Table 1-2

Partnership Flying Expenses
Used Four-Seat Single

Number of pilots		**1**	**2**	**3**	**4**
Purchase price per pilot		**45,000.00**	**22,500.00**	**15,000.00**	**11,250.00**
Annual fixed expenses					
Tiedown/hangar		900.00	450.00	300.00	225.00
Insurance		1,500.00	750.00	500.00	375.00
State fees		120.00	60.00	40.00	30.00
Annual		750.00	375.00	250.00	187.50
Maintenance		1,000.00	500.00	333.33	250.00
Loan payments		3,443.30	1,721.65	1,147.77	860.83
Cost of capital (noncash)		*1,750.00*	*875.00*	*583.33*	*437.50*
Total fixed expenses/year		**9,463.30**	**4,731.65**	**3,154.43**	**2,365.83**
Hourly operating expenses					
Fuel		20.00	20.00	20.00	20.00
Oil		0.13	0.13	0.13	0.13
Engine reserve		5.00	5.00	5.00	5.00
General maint res		2.00	2.00	2.00	2.00
Total operating expenses/hour		27.13	27.13	27.13	27.13
Total hourly expenses (fixed+op)					
50 hours		216.39	121.76	**90.21**	74.44
100 hours		121.76	**74.44**	58.67	50.78
150 hours		90.21	58.67	48.15	42.90
Hourly commercial rental		90.00	90.00	90.00	90.00
Total annual expenses per pilot					
50 hours	Own	1,0819.55	6,087.90	4,510.68	3,722.08
	Own (cash only)	9,069.55	5,212.90	3,927.35	3,284.58
	Rent	4,500.00	4,500.00	4,500.00	4,500.00
100 hours	Own	12,175.80	7,444.15	5,866.93	5,078.33
	Own (cash only)	10,425.80	6,569.15	5,283.60	4,640.83
	Rent	9,000.00	9,000.00	9,000.00	9,000.00
150 hours	Own	13,532.05	8,800.40	7,223.18	6,434.58
	Own (cash only)	11,782.05	7,925.40	6,639.85	5,997.08
	Rent	13,500.00	13,500.00	13,500.00	13,500.00

Assumptions		
	Aircraft value ($):	**45,000.00**
	Loan O/S ($):	**20,000.00**
	Loan interest rate (%):	12.00
	Loan/inv term (yrs):	10.00
	Cost of capital rate (%):	7.00
	Hangar/month ($):	75.00
	Insurance/yr ($):	1,500.00
	State fees/yr ($):	120.00
	Annual ($):	750.00
	Maintenance/yr ($):	1,000.00
	Fuel consumption (gal/hr):	10.00

Continued Table 1-2

Assumptions	(Continued)	
	Fuel cost ($/gal):	2.00
	Oil consumption (qt/hr):	0.13
	Oil cost ($/qt):	1.00
	Gen maint res/hr ($):	2.00
	Time before overhaul:	2,000.00
	Engine major OH. cost ($):	10,000.00
	Commercial rental/hr ($):	90.00

Additionally, the cost of capital (noncash) is the amount you are giving up in income by buying an airplane instead of investing your money in an income producing investment such as bonds or mutual funds. It is not a cash outflow from your pocket because it is not money you are paying out, it is money that you are not receiving. If you only want to know how much out-of-pocket cash your flying is going to cost every year, you can ignore this category (see "cash only" annual expenses instead).

The rest of the table is self-explanatory. If you wish to get deeper into the advantages and disadvantages of partnerships (and there are many), see *Fly for Less: Flying Clubs & Aircraft Partnerships* by Geza Szurovy, McGraw-Hill 1992.

21 Performance per dollar

* **Action** Purchase the airplane that delivers the amount of performance you want for the lowest purchase price. There is a vast army of used production airplanes out there with decades of variance in age. It is a fact that older models of the same type tend to cost considerably less than newer models. Yet well-maintained older models of the same type generally have engines that are much younger than the airframe and deliver performance equal to the newer models. What can a 1989 Piper Arrow with a fresh engine do that can't be done just as well by a 1978 Piper Arrow with a fresh engine? The airframes are identical and so are the engines, but the '78 Arrow sells for thousands of dollars less, even with new paint and interior.

Nor do you have to stick to the same type. The fact is that there have been practically no widely adopted advances in production aircraft airframe and engine design for the last 30 years. Any basically good general aviation airplane designed for a particular mission with a similar engine will perform within the same ballpark regardless of age and type. A properly maintained and periodically refurbished 1968 Skyhawk will do just as well for half the price as a 1983 Warrior. And the longevity of Bonanzas is legendary. A refurbished 1950s Bonanza with a fresh engine gives the latest model a run for its money and looks practically identical for a fraction of the price.

To get the performance you want for the lowest price you have to really do your homework. Carefully research the performance criteria of all potential options. In examining the various types, go as far back as the first year of production. Sometimes significant reconfigurations took place during various model years for a particular type. The most common change over time is the introduction of engines with greater horsepower; for example, Cessna 172s manufactured through 1967 were made with only the 145-hp Continental that had a TBO of only 1,800 hours.

Carefully check to see if any airframe or engine modifications are available that effectively convert an early model aircraft into a later model, or very close. Early 150-hp Piper Warriors, for example, can be upgraded to the later 160-hp model. This modification doesn't come close to increasing the price to the level of a later model Warrior but the performance is matched after the changes.

Research AD histories and any particularly troublesome model years that could reduce the value of an apparent bargain. Be especially wary of any expensive recurrent maintenance items particular only to certain model years of a type. Many enduring older aircraft have a fanatically loyal following that is usually organized into very well informed type clubs. Unless you know someone with all the answers, the type club is the best place to go when you have a question. (*See* Resources in the back of the book for type-club addresses.)

If you go the older airframe route, spare no effort in the mechanical check. A well-maintained, periodically refurbished older model in compliance with all ADs should be no more expensive to maintain

than newer models, but a mechanically neglected old airplane can be a nightmare. As a rule, avoid engines that have been overhauled more than once and don't touch any aircraft without a complete set of airframe and engine logs from date of manufacture, unless it is a thoroughly documented ground-up restoration (in which case the restoration date becomes day one).

If you wish, the approach of buying the same performance for the least amount of dollars can be varied to get the *most performance* for a fixed amount of dollars. This subtle difference in approach might enable you to pick up a real hot rod of an old Mooney that leaves equally priced but much less capable types of newer vintage far behind.

$ **Savings** Scrutinizing performance per dollar could save tens of thousands of dollars. Tables 1-3 and 1-4 indicate the vast price differences of aircraft with similar performance but different age. (The averaged prices were listed in the N.A.D.A. *Retail Aircraft Appraisal Guide* at the time of writing.)

Table 1-3 shows the prices of Cessna Skyhawks for different model years. Note the differences in engine power: 160 hp for the latest models and 145 hp for the earliest models, as well as some airframe differences in the early models. The associated performance differences are fairly minor in everyday use.

Table 1-3

Aircraft type & model year	Price ($)
1986 Skyhawk	79,300
1983 Skyhawk	65,900
1980 Skyhawk	40,900
1977 Skyhawk	33,400
1974 Skyhawk	29,500
1969 Skyhawk	24,900
1962 Skyhawk	20,000
1956 Skyhawk	16,100

Table 1-4 shows different aircraft types and different model years. All aircraft are 180-hp, fixed-gear four-seat singles. In spite of the tremendous price difference, they all deliver essentially the same performance.

Table 1-4

Aircraft type & model year	Price ($)
1988 Aerospatiale Tobago	126,700
1984 Beech Sundowner	59,400
1980 Piper Archer II	54,300
1978 American Tiger	40,400
1974 Piper Challenger	36,200
1968 Piper Cherokee 180	27,100

Registration strategies

Action Seek a legitimate way to register your aircraft in a state with the lowest registration fees. There are significant differences in the registration fees charged by states. If you live in the border area of two states that charge different fees, you might find a way to register the airplane in the state with the lower fee. Check out the registration criteria and see what you can do. Normally the airplane has to be based in the state of registration and the owner of record has to be a resident of the state. The key is what constitutes "residence."

In case of a partnership in which the partners live in two different states, the airplane could be registered to the partner who is a resident of the state with the lower registration fee.

$ **Savings** Careful registration could save up to approximately $100 per year. Check out registration fees and requirements with the state aviation authorities. Call your state house, licensing and registration fees section for information. (*See* Resources in the back of the book.)

Rental vs. ownership

✳ **Action** Ownership is an expensive proposition. You have to fly a minimum number of hours per year to make it less expensive than rental. Before you decide to take the ownership plunge, you should calculate the break-even point in comparison to rental and see if you are likely to fly the number of hours necessary to make ownership financially more attractive.

Here is a simple way to figure out which option makes the most financial sense. First, collect the relevant data:

- ➢ FBO or flying club rental per hour: $50
- ➢ Annual fixed costs of ownership: $5,000 (tiedown, insurance, annual, payments, etc.)
- ➢ Hourly operating cost of owned aircraft: $20 (fuel, oil, engine and maintenance reserve)

Next, perform the required calculations:

- ➢ Annual cost of flying at the FBO for 25, 50, 75, 100, 125 and 150 hours: $50 × hours flown per year.
- ➢ Annual cost of owning for 25, 50, 75, 100, 125 and 150 hours: Hourly operating cost ($20) × hours flown per year + annual fixed cost of ownership ($5,000).

Table 1-5 is the results in a table that reveals your answer. As you can see, break-even is somewhere between 100 to 125 hours. If you fly 125 hours per year as an owner, you will be $625 ahead of the rental option. If you want to be more precise, you can extrapolate. Take 112 hours (the halfway point between 100 hours and 125 hours) and calculate the cost: rental is $7,280; ownership is $7,250. So, if you fly fewer than 112 hours per year, you are financially better off renting; if you fly more, it makes financial sense to buy.

Table 1-5

Hours per year	Rent per year ($)	Own per year ($)
25	1,625	5,500
50	3,250	6,000
75	4,875	6,500
100	6,500	7,000
125	**8,125**	**7,500**
150	9,750	8,000

Table 1-6 is a chart that quickly shows break-even. The vertical axis is dollars, the horizontal axis is hours. All you need to do for rental and ownership, respectively, is calculate the cost of 25 hours and 150 hours, plot these points on the graph and connect them. The break-even point is where the ownership and rental lines cross. Read off the break-even hours and dollars on the respective axes.

Table 1-6 **Rental vs. Ownership Break-even Analysis**

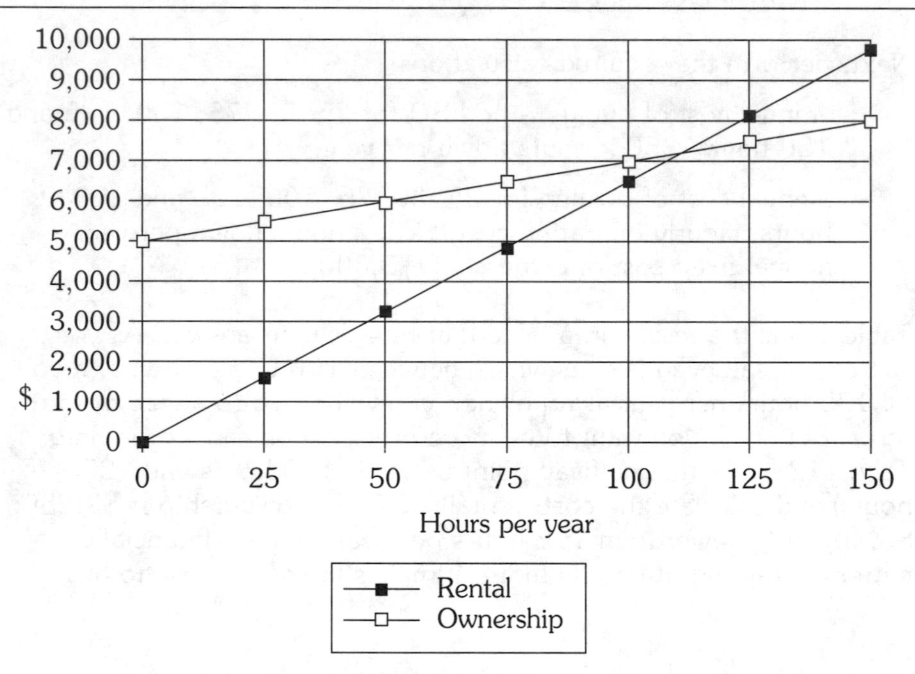

Bear in mind that your analysis will be only as accurate as the data on which you base it. Set realistic costs and annual flight hours. Tip #20 in this chapter presents a detailed model to assess ownership costs.

$ **Savings** Analyzing the rental or ownership options could save you thousands of dollars. This type of analysis can save you substantial amounts of money in the sense that it shows you the minimum number of hours you must fly to financially justify ownership. Many people who own aircraft don't fly nearly the number of hours to warrant ownership. They could save hundreds or thousands by renting instead.

If your analysis does not justify single ownership don't despair. Before you decide to rent instead, consider a partnership (as described in this chapter) and rework the numbers with different numbers of partners.

24 Sell abroad

Action Sell your airplane abroad for a higher price than what you can get in the United States. There are times when prices for used aircraft are higher overseas than domestically. This is especially true when the dollar is weak. If the dollar is weak, the same amount of foreign currency buys more dollars than when the dollar is strong (Fig. 1-3).

Figure 1-3

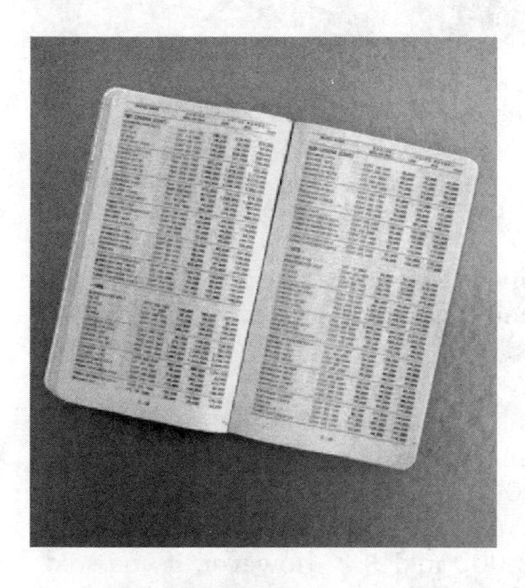

Be sure to know how much to ask when you are selling your airplane.

Regardless of the value of the dollar, the typical foreign buyer is willing to pay the same amount of foreign currency for a particular airplane type than he is used to paying; when the dollar is weak, that same amount of foreign currency is worth more dollars than when the dollar is strong. You, the seller, get more dollars, even though the buyer pays the same foreign currency price that has always been paid (Fig. 1-4).

Figure 1-4

This Cutlass was sold to a French pilot for much more than the owner could have hoped to get in the United States at the time.

Aircraft brokers who have caught on to this idea make a lot of money. They buy the airplanes from United States sellers for the domestic price and sell the airplanes to foreigners for the typical price abroad. The brokers clean up when the foreign currency is converted into dollars.

For example, one dollar was equal to one pound sterling (£, the British currency) in 1982; a Cessna 150 was going for £12,000 in 1982, which was equal to $12,000. By 1988, the dollar had weakened considerably; $2 were now equal to £1. The Cessna 150 was still selling in Britain for £12,000 in 1988; however, if you sold one and converted the pounds into dollars, you now got $24,000. Even with a $5,000 shipping cost, it was worth the trouble in 1988 to sell Cessna 150s to Britain.

The question is how to contact foreign buyers directly. Perhaps the best way is to advertise in foreign aviation magazines. American bookstores specializing in hard-to-find magazines sell the English language foreign flying magazines; the EAA library carries practically all foreign flying magazines published in any language. Find a translator to help you place an ad. Find out from a United States shipping broker (they all advertise in *Trade-A-Plane*) what the shipping cost would be to the magazine's country, look up in the magazine ads what the local average price is for the type of aircraft that you are selling, and lowball your asking price for quickest results. When the ad appears, you might be contacted by a private buyer as well as a foreign broker.

Work through the United States shipping broker and its recommended bank to complete the paperwork and assure that you get paid. Arrange to receive payment through a dollar-denominated letter of credit to avoid any risk of nonpayment and currency fluctuation.

$ **Savings** Hundreds or thousands of dollars are saved "in reverse" by getting more money for your airplane overseas than you would get domestically. Note that this savings tip is wholly dependant on the exchange rate for the dollar. It is entirely possible that the dollar will at some point strengthen so much (it has happened in the past) that it will become more attractive to buy rather than sell used aircraft abroad. In that case, the tip still works, but in reverse.

25 Single vs. twin

✻ **Action** Select a single-engine airplane with equal performance of a twin. Certain higher performance singles match the performance of some light twins in speed and range, yet offer considerable savings in three areas: acquisition, maintenance, and operations.

Regarding acquisition, a single-engine aircraft can often be less expensive to buy than an equivalent twin.

Regarding maintenance, savings in maintenance costs can be especially attractive. The fee for the annual inspection of a single-engine airplane is lower and maintaining that engine and its associated accessories is always less expensive than having to foot the bills for the upkeep of two engines.

And lastly, regarding operations, avgas and oil for one is always less expensive than two.

The safety of twin-engine redundancy is always a big attraction, but this perception is not necessarily upheld by accident statistics that involve an engine failure of a twin-engine airplane.

$ **Savings** Flying a single-engine airplane instead of a multi-engine airplane will typically save several hundreds of dollars per year; you might save thousands of dollars. Annual savings of $500 to $2,000 are common in annual fees and maintenance. The cost of having to overhaul only one engine when the time comes is a major maintenance saving. Depending on annual flying hours and fuel consumption, savings on annual fuel bills can also amount to thousands.

Compare the performance of likely aircraft types carefully to identify singles that measure up to the important performance criteria of the twin you favor. Check the price books to identify the best bargains. Here are a few single/twin examples from the N.A.D.A. book:

1963: Comanche $37,600/Twin Comanche $41,000

1978: Lance $63,000/Seneca $73,500

1982: A-36 Bonanza $156,000/B-55 Baron $171,000

26 State sales tax strategies

* **Action** State sales taxes vary significantly. Find a way to be liable for sales tax in a low-tax state, or better still, register the sale in a state where there is no sales tax. Residence is the usual test for sales tax liability. Research the criteria for "residence." Pilots living in the border areas of states with big differences in sales tax might be in the best position to benefit from this opportunity.

$ **Savings** Hundreds to thousands of dollars might be saved in taxes. Given that the typical sales tax is in the 5–7 percent range, savings can be substantial.

27 Tiedown costs

✻ **Action** Scout around at your area airports for the least expensive tiedown rates within reasonable commuting distance. Rates fall rapidly as soon as you get only a few miles outside larger urban areas.

$ **Savings** Expect to save $20–$40 per month ($240–$480 annually). You might find as much as a $40 difference between tiedowns at large general aviation airports in metropolitan areas and smaller airfields less than an hour's drive away. The $480 in annual savings might be well worth it to the budget-conscious flyer if the savings are not burned up in gas expenses for a commute; be sure to factor in automobile expenses if you are considering driving a long distance to an airport.

2

Renting an aircraft

28 Aircraft choice—most bang for the buck

✻ **Action** Whatever your rental source, rent the aircraft that meets your mission criteria for the lowest hourly rate. Aircraft design has not progressed much in the last 30 years. All aircraft produced in the last 30 years to fulfill the same mission, equipped with the same engine, will turn in the same performance. This includes a vast pool of older, mechanically well maintained aircraft that are considerably less expensive to rent than newer models. An early model no-frills 1977 Skyhawk will fly as well as a 1987 IFR Warrior. Both have fixed gear, a fixed-pitch prop, four seats, and a 160-hp engine (generally overhauled or replaced every 2,000 hours). Both will carry four occupants with equal performance on a VFR flight, yet the Skyhawk will most likely rent for about $20 less.

$ **Savings** A $10–$30 per hour savings is possible, depending on various factors.

29 Bartering for flying time

✻ **Action** If you have any services to offer to owners of aircraft available for rent, barter your services for flying time. Establish a value for your services and arrange for a matching value of flying time. The key to savings through barter is in you performing the services and the owners providing the flying at cost. The airplane owner gets a service at a rate below what it would cost otherwise, and the provider of the service gets to fly at a preferential rate. Typical examples of services bartered for flying time are a bookkeeper's services to an FBO, a mechanic's services to a private owner, legal services to a flying club, and so on.

$ **Savings** Hundreds of dollars or more can be saved by both parties in the barter transaction.

30 FBO—block time discounts

✻ **Action** Take advantage of block time discounts offered by most FBOs. If you pay in advance for a block of time (say, 50 or 100 hours) you can get a break on the rate. The more time you buy, the greater the discount.

A potential problem with paying in advance might be retrieving your unused money in case the FBO goes out of business. It is difficult to reliably evaluate the solvency of an FBO, but you can take some comfort from past history, the consistency of management, and the level of flight activity. Keep your eyes and ears open for any sudden changes, and don't go overboard buying a lot of block time at once.

$ **Savings** Up to 20 percent of regular hourly rates might be saved. If the regular rate for an hour is $75, a 15 percent block time discount for 50 hours yields a savings of $1,125.

FBO—choosing the right FBO

Action Rent from the FBO in your area that has the best rates for good airplanes. Rental rates at FBOs are by no means uniform. Many factors go into setting rates. Among them are the cost of premises, the cost of financing on the fleet and any other loans, the cost of avgas to the FBO, and so on. Overhead at the smaller, more outlying airports is generally lower than the big metropolitan general aviation airports.

Some older FBOs might have loans with lower interest rates on the fleet than newer operations, or the older FBO might own the fleet outright. An older fleet of less glamorous two-seat trainers might have lower carrying costs than a newer fleet equipped with the latest four-seat tourer/trainers.

FBOs at outlying smaller airports might be owner/operators of the airport with associated financial advantages. Other FBOs might have excellent lease terms from a local community keen to have a community airport (it does still happen). All these elements and others like them influence the rates an FBO needs to set for the fleet. It behooves you to shop around for the best deal.

$ **Savings** Typically $5–$25 per hour can be saved, depending on type and whether solo or dual. One FBO has Cessna 150s for $45 per hour. Another FBO instructs in Warriors for $70 per hour. You save $25 per hour flying the Cessna 150s (Fig. 2-1). If you rent an airplane for 50 hours a year, your annual savings for the rental will be $1,250. You get the idea.

FBO—flying club rates

Action Take advantage of the "flying club" rates most FBOs offer. These rates are usually 10–20 percent below regular rates. To get the rates, you usually have to pay a so-called *annual flying club fee* up front. Factoring in this fee, you have to fly a certain minimum number of hours before you are ahead compared to the regular rates.

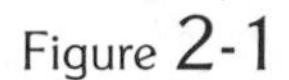

Figure 2-1

A budget Cessna 150 will teach you to fly for a lot less than popular fancy alternatives.

From the FBO's standpoint, these rates are designed to increase fleet use by encouraging renters to fly the minimum number of hours per month to benefit. Typically the fee is set so that if you fly 2–3 hours per month, it is worth your while to get the flying club rates. Here is an example of how the math works:

Regular rate per hour: $55

FBO "flying club" rate: $45

Hourly "flying club" savings: $10

Annual "flying club" fee: $250

Hours/year at club rates to break even with regular rates:

$250 \div 10 = 25$

Hours/month to break even: $25 \div 12 = 2$

FBOs have several variations on this scheme for flying club rates. Have them explain the details and take you through the math.

$ **Savings** Typically 10–20 percent saved, over regular rates. When the minimum hours are flown at club rates to reach break-even with regular rates, you benefit from the entire difference between the two rates. The example shows that when you have flown 25 hours, you save $10 per hour. At 100 hours, for example, that is a savings of $1,000.

33 FBO vs. flying club

* **Action** Renting at a flying club independently of an FBO might seem to be always less expensive than renting from an FBO, but this is not always the case. While the hourly rental rate for comparable aircraft is invariably lower, the hefty annual dues of the typical flying club must be taken into account. The annual dues must be spread over the number of hours you plan to fly and added to the hourly rental rate. You will find that below a certain number of hours you are better off at the FBO. Here is how to do the math (which is the same calculation used to consider the FBO's flying club rate):

FBO rate per hour: $55

Flying club rate per hour: $35

Hourly flying club savings: $20

Annual flying club fee: $500

Hours/year at club rates to break even with FBO rates: 500 ÷ 20 = 25

$ **Savings** Typically hundreds or thousands of dollars per year might be saved depending on rates, dues, and hours flown. In the example, you have to fly 25 hours to be even with flying club rates. You will be realizing savings per every hour beyond that. Compare the cost of 100 hours of flying per year at the FBO and at the flying club. The FBO cost is 100 × 55 = $5,500. The flying club cost is: 100 × 35 + 500 = $4,000. The annual savings at the flying club is $1,500.

34 Flying club—choosing a flying club

✻ **Action** Choose the flying club in your area that charges the lowest fees yet meets your flying needs. Rates vary greatly among flying clubs depending on purpose, number of members, type of aircraft, number of aircraft, and many other factors. As you evaluate flying clubs, keep some points in mind with a financial perspective: purpose, aircraft, and dues and rates.

Regarding purpose, some clubs are lavish organizations, designed to provide high-quality expensive flying with the least amount of hassle. Others are established explicitly to provide budget flying.

Regarding aircraft, the aircraft and its cost will reflect the club's purpose. Look for a club with aircraft that meet your performance requirements at the lowest cost. A club equipped with two 1960s vintage Cherokee 180s is most likely a better bargain than club flying two 1980s vintage Archer IIs, yet both aircraft deliver basically the same performance.

And regarding, dues and rates, the easiest way to bargain hunt is to check the dues and hourly rental rates. Most clubs will give them to you in an information package. See if you qualify for any preferential discounts, such as reduced student dues.

See how high the annual dues are and how much they will add to your hourly rate given the number of hours you plan to fly:

Annual dues: $500

Hourly rental: $45

Planned hours per year: 100

Hourly effect of dues: 500 ÷ 100 = $5

Total hourly rate: 45 + 5 = $50

Perform this calculation for the dues and rates structure of all the clubs you are evaluating.

Perhaps you could work for flying opportunities. Some clubs might have a need for services for which they might be willing to offer free flying or flying at a reduced rate. These services might be aircraft related such as mechanical work if you are an A&P, or not related to flying at all, such as bookkeeping or accounting.

Explore block time discounts. Check to see if the club has any block time discount program.

$ **Savings** Joining a flying club might save you hundreds of dollars. Consider this example. If you find a flying club where you save only $5 an hour compared to other clubs and you fly 100 hours a year, your savings will be $500 per year. You can use it for groceries or more flying. If, for example, your hourly rental rate is $30, you can fly an additional 16.7 hours each year.

35 Nonprofit rental from private owners

✻ **Action** Rent from a private owner on a nonprofit basis at rates below FBO and flying club rates. This is a touchy issue because of concern about such rentals being commercial operations. Under the FARs, if the private owner receives compensation for expenses but not a profit, the venture is not a commercial operation. This practice is not widespread, but from time to time private owners might find it advantageous to rent their aircraft to individuals to defray the cost aircraft ownership. If you are in a position to enter into such an arrangement you can significantly cut your hourly flying costs.

Warning: Make sure that the aircraft is properly insured for you to act as pilot in command; the private owner might not be able to let you fly her airplane until you have been formally placed on the insurance policy. The more experience you have— instrument rating, commercial certificate, and the like—the better likelihood that you can be easily placed on the insurance policy. Regardless of your experience, shop for nonowner's (renter's) insurance that protects you if the underwriter for an FBO, flying club, or private owner sues you for damages to the aircraft.

$ **Savings** It is common to save $15–$35 per flying hour, based on a variety of factors.

3

Accessories, parts, and supplies

36 Aviator sunglasses

✻ **Action** Buy your aviator sunglasses at a discount apparel store rather than an aviation specialty shop, and buy them on sale. Aviator sunglasses are chronically cool. Millions who will never see the inside of a cockpit can't imagine being without their Ray Bans.

The stores catering to them buy the shades in such high volumes that an aviation store can rarely if ever compete with the discounters. Check the airport pilot shop or the latest aviation goods catalog for what's most "in" for flying types, and head to the discount store to find it.

$ **Savings** Savings of as much as $40 are not unusual on the more expensive glasses.

Avionics—buying

✱ **Action** Buy avionics from discount suppliers (mail-order houses are the most reasonable) rather than the local radio shop. It is the same old story. A store that can order 25 navcoms at a time will get a big fat discount off the unit price compared to what the shop that buys two at a time has to pay. You might use the discount supplier's quote to bargain with the local shop. In rare instances, a match might be possible.

Note that when you buy an avionics item from a discount store, you have to get it installed somewhere, most likely a local radio shop. The installation charge alone will be heftier than the installation charge portion of avionics bought at the local shop; however, in most instances the total costs still justify purchasing from the discounter.

A word of warning: Don't confuse overhauled (often referred to as "yellow tagged") avionics with factory-new units. Tip 41 in this chapter examines new and overhauled avionics.

See publications like *Trade-A-Plane* to identify mail-order discounters. Whatever you do, shop around and bargain blatantly.

$ **Savings** You will find that 15–30 percent savings over local shop prices are not unusual, which can amount to hundreds of dollars.

Avionics—how much equipment is too much?

✱ **Action** Don't overequip your airplane. An HSI, or even just dual VORs are excessive in an airplane used only for Sunday coffee shop tours in CAVU weather. If you fly only VFR on a budget, stick to bare bones VFR avionics (Fig. 3-1).

Figure 3-1

This panel is adequate for the Sunday-morning CAVU coffee shop rounds.

Honestly assessing your needs is the key to making the right decision about how much avionics to buy. More pilots than you think end up shelling out thousands of unnecessary dollars on avionics because they plan all sorts of long distance adventures and ratings that never materialize (Fig. 3-2).

Figure 3-2

This panel will get you around the world, but is a waste of money if you are not going anywhere.

Every year hundreds of airplanes set out for Oshkosh from hundreds or thousands of miles away and get there on a wet compass. It can be done and will even make you a better pilot. That is not to say you shouldn't get a budget GPS to simplify life; just don't overdo it.

$ **Savings** Careful buying will save thousands of dollars. Consider these avionics packages offered in a recent issue of *Trade-A-Plane* (both packages are brand-new avionics, installation is extra): basic VFR package, 1 navcom, transponder, and audio panel for $3,125; basic IFR package, 2 navcoms (one with glide slope), 1 ADF, transponder, and audio panel for $9,350.

The VFR package savings is $6,225. Most VFR pilots would find this VFR package—plus a yoke-mounted GPS/moving map for another $1,500—all the avionics they will ever need (a total cost of $4,625 with the GPS).

39 Avionics—installation

✳ **Action** There are two possibilities for saving on avionics installation. One, find a freelancing appropriately qualified avionics technician. Ask around. Big airfields with a lot of avionics maintenance activity are good places to explore. A&P schools that provide avionics training and employ avionics instructors are another good source.

Two, if you are buying avionics from an area nondiscount source, always have the electronics installed at the same place you buy them. Some of the discount mail-order suppliers are also avionics shops and can match or beat the local shop's installation rates.

$ **Savings** Savings of several hundred dollars might be possible depending on the amount of labor involved. Savings will be realized on labor. The free-lance avionics technician will charge well below shop rates to be competitive. If you get the avionics installed by the same place where you bought them, you can expect a discount on installation of 10 percent or more.

Avionics—kits

⁂ **Action** Buy an avionics kit and assemble it yourself. From time to time manufacturers offer peripheral avionics that can be bought in kit form. Anyone comfortable with a soldering gun can assemble the kit, send the completed item back to the factory for calibration if required, and be in business. Not only is this option less expensive than buying the finished product, it is also great fun.

Don't expect to assemble a 720-channel navcom in your basement. Important mainstream avionics are not available in this fashion; however, intercom kits as well as simple radio kits for use in sailplanes and ultralights have enjoyed great popularity from time to time. Keep your eyes open for kits and check out the "you do it" electronics stores.

$ **Savings** A kit price might be 50 percent off the purchase price of a similar preassembled unit.

Avionics—new vs. overhauled

⁂ **Action** Purchase overhauled avionics instead of factory new avionics. Some mainstream avionics items, such as King's popular KX155 series navcom have been around for many years. Over time a good secondary market has developed in overhauled units, which are considerably less expensive than brand-new units (Fig. 3-3).

Is there a risk in getting an overhauled unit? Not really, if it is from a reputable overhaul facility. Overhaul in the world of today's solid-state electronics consists mostly of the simple replacement of offending parts with new components. Cosmetic refurbishing and extensive testing and calibration completes the job and the overhauled item is practically as good as new. These units are often referred to as *yellow-tagged* units because the overhaul is always documented on a yellow tag.

Figure 3-3

Reconditioned avionics dramatically cut the cost of this new panel.

$ **Savings** Several hundred dollars per unit can be saved. To get a good idea of savings opportunities, consider these price comparisons for a popular modern navcom.

	Factory new	Overhauled
Nav/Com/VOR/GS	$2,737	$2,250
Nav/Com/VOR	$2,350	$1,850

42 Avionics—upgraded radios

* **Action** Exchange your once popular but now obsolete King KX170 series, Narco Com 11/Com120 or various ARC navcoms for highly modified, versions of the same unit updated with the latest digital technology and flip/flop displays.

Two modification manufacturers, TKM and McCoy, offer such radios, which are straight replacements. No rewiring or remounting is necessary. Just slip out the old radio, slip in the replacement, and you

are in business for much less than it would cost to buy a navcom of more recent design, overhauled or factory new.

$ **Savings** Expect to save more than a thousand dollars per radio. Savings are realized not only on a substantially lower purchase price, but also on the lack of installation costs. A replacement for a KX170B cost $1,295 with an exchange. An overhauled KX155 including installation, which requires substantial wiring and mounting work, costs $2,800, less $400 for the trade in of the old KX170B, or a total of $2,400. The savings in this instance is $1,105.

Brakes—brake pad purchase

✻ **Action** Brake pads are made from two different materials, nonmetallic material and more durable metallic material. Buy brake pads made from the long-lasting, more durable metallic material. The tougher brake pads are more expensive than the regular ones, but the durability exceeds the extra cost, so they are the better buy.

$ **Savings** Expect about a 10–20 percent savings comparison to regular brake pads. The gains are recognized because the more expensive durable pads last longer, eventually offsetting the cost difference.

Brakes—chromed discs

✻ **Action** Get chromed discs if you need to get new brake discs. This is a long-term savings proposition. Chromed discs actually cost somewhat more than steel discs, but they will last as long as the airplane (Fig. 3-4). Regular steel discs, on the other hand will have to be replaced again at some point due to wear and tear.

This option is especially suited for airplanes that don't fly much. Standing idle for long periods of time exposes brake discs to the elements, greatly increasing the chance of rust building up on regular brake discs, which will result in pitting. Chromed discs are rustproof.

Figure 3-4

Chromed discs pay off in the long run.

$ **Savings** Over time, savings can be several hundred dollars due to the elimination of the need for additional periodic brake disc changes.

45 Budget headset

✻ **Action** Make an inexpensive earphone instead of buying an expensive headset. Buy a portable cassette headset or a single earphone. Chances are that you already have one lying around from one of the many portable stereo gadgets you have around. Buy a headphone adapter plug that allows you to use the earphone with the large diameter headphone jacks found in aircraft. Plug the earphone into the adapter and plug the adapter into the headphone jack in your airplane (Fig. 3-5).

If you use a portable cassette stereo headset, be sure to buy a stereo to mono adapter (stereo headset going into mono jack); otherwise you will hear the radio only in one ear.

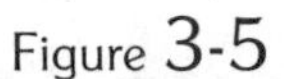

Figure 3-5

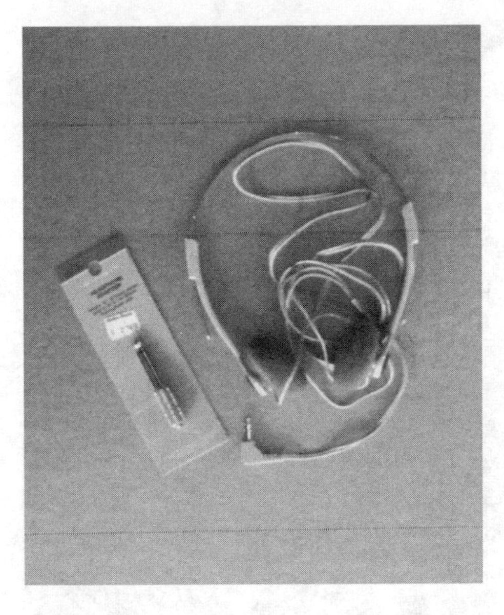

An innovative budget alternative to a headset. The earphones were free on an airline, and the adapter plug cost under three dollars at an electronics store.

If you get the single earphone, also get a plastic holding device that keeps it in place by hanging on your ear, or fashion one from a piece of plastic coated wire.

This inexpensive solution fulfills the primary purpose of earphones. You hear radio transmissions much more clearly than without earphones. You still have to use the hand-held microphone to transmit and you don't get the noise protection offered by the massive padded ear cups of the commercial headphones. But you also reduce headphone costs by a factor of 10 to 15. The inexpensive headphones are also great as backup headphones and passenger headphones.

$ **Savings** You will pocket $100–$200 per headphone. The inexpensive headphones can be assembled for under $20, even if you have to buy all the pieces specifically for this purpose. The adapters cost under $3 in places like Radio Shack. By comparison, pilot's headphones start at around $120.

Budget IFR hood

Action Make your own set of view-limiting plastic glasses instead of purchasing the commercially available alternative. Get a set of clear safety glasses. Mark the area you want to remain clear. With medium grit sandpaper sand the rest of the viewing area until it becomes opaque (Fig. 3-6). Wash and rinse the glasses several times to make sure the plastic residue is cleaned off.

Figure 3-6

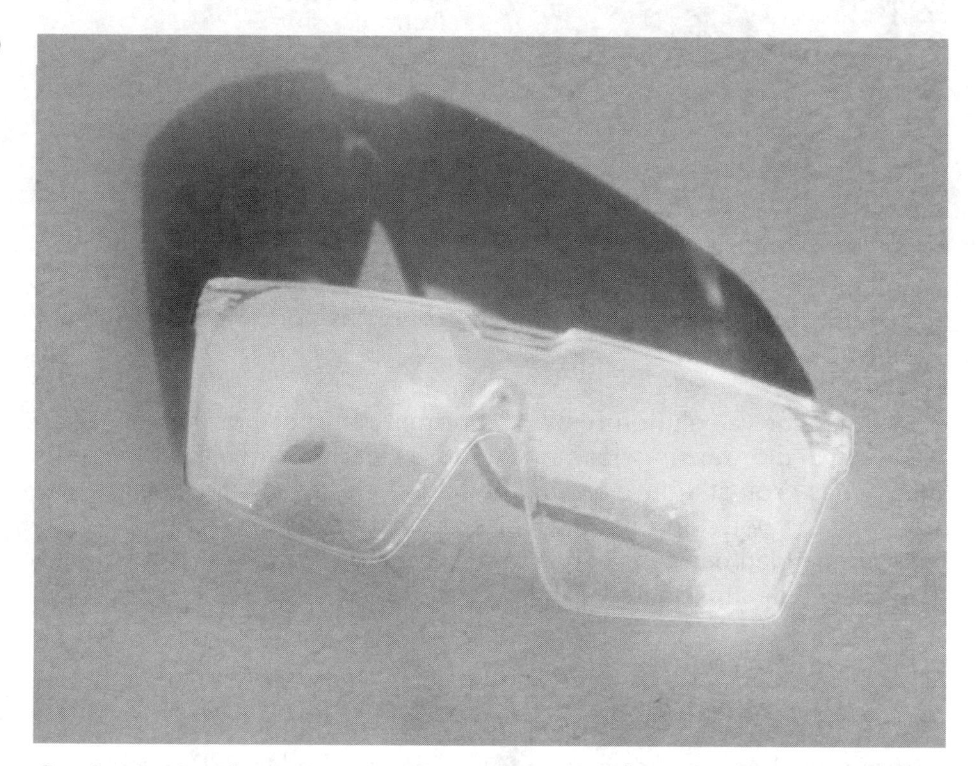

Sandpaper (medium, then fine) converted these $3.50 safety glasses into an IFR view limiting device in 5 minutes. Thoroughly wash and dry the glasses several times to ensure removal of all plastic particles.

Savings You will save \$10–\$15. Clear safety glasses cost around \$3.75, much less than the \$18 being asked for view-limiting glasses.

47 Budget oil funnel

Action Make an oil funnel out of a clean disposable plastic soft drink bottle. The 33.8-oz. and 12-oz. bottles are the most suitable. Larger bottles are too unwieldy. Turn the soft drink bottle upside down. Cut off the bottom with a pair of scissors. To penetrate the plastic safely, prior to cutting off the bottom, saw it gently with a hacksaw, or make a small incision with a very sharp utility razor knife. You will find that the screw top of some bottles perfectly screws into the dipstick receptacles of some Lycoming engines (Fig. 3-7 and Fig. 3-8).

Figure 3-7

A pair of scissors and a clean, empty plastic soda bottle . . .

$ **Savings** You will save $5–$10. The plastic bottles are free. You save the cost of the funnel.

Figure 3-8

. . . Quickly fabricate a free instant oil funnel.

48 Budget yoke-mounted clip

✻ **Action** Make an inexpensive yoke-mounted clip for approach plates and checklists. Get a broom-handle holder (that mounts on the wall or inside a door), a fully threaded bolt of suitable length (3 inches seems to work well) to place the clip as high above the yoke as you want it, a paper clamp, and six nuts (Fig. 3-9).

Slip the bolt through the broom-handle holder, and secure it with two nuts, one on top of the other to prevent unscrewing. Bolt the paper clamp to the other end of the bolt (two nuts on each side of the clamp). Snap the broom-handle holder in place on the yoke (Fig. 3-10).

Before you buy the broom-handle holder, make sure that the tube on your yoke is the right size to accommodate it. Otherwise you might have to scrounge around for some alternative clip-on device.

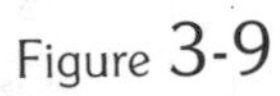
Figure 3-9

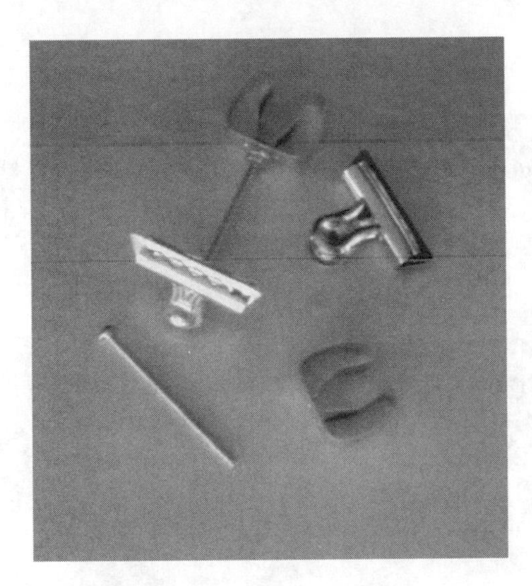

A handful of discount hardware and stationery store items make an instant budget yoke clip.

Figure 3-10

The homemade yoke clip is instantly mounted in practically any aircraft.

$ **Savings** You will save \$5–\$10 and more. The savings depends on how much of the material you can scrounge and how much the alternative commercial products cost. A preassembled version of the described clip is available for between \$5–\$10.

49 Canopy cleaner

✳ **Action** Clean your canopy with toothpaste very slightly diluted in water. Toothpaste is much less expensive than mirror glaze and works just as well. Dilute the toothpaste so that it is less dense, but still a paste. Get plain old-fashioned toothpaste. The canopy doesn't care about any fancy additives. Apply the paste, let it dry, and rub off vigorously with a soft cloth, rubbing up and down rather than in a circular motion.

$ **Savings** You will save \$5–\$10. Savings are realized not only because a tube of toothpaste is less expensive than mirror glaze, but also because a large tube of toothpaste lasts as long as two to three bottles of mirror glaze.

50 Charts—sourcing

✳ **Action** There are several ways to save on charts. Get your charts from a discount source. For IFR charts, use slightly less expensive government NOS charts instead of the alternative provided by the private sector. Use WAC charts instead of sectionals if you go on long cross-country flights extending beyond the area covered by a single chart. You will need fewer of them and will find your way with them just as easily.

$ **Savings** You will save 10 percent or more compared to the retail alternative, if you buy from a discount source. NOS charts are also good for a 10 percent savings. WAC charts might cut your chart expenses in half in comparison to sectionals.

51 Charts—type

✻ **Action** Get only the charts appropriate to the kind of flying you do. Don't go overboard subscribing to IFR charts if you don't keep current. If you fly IFR, subscribe only to coverage of the area in which you really fly. You can end up spending a lot of money on charts unnecessarily if you don't have a realistic idea of the type of flying you are likely to do.

$ **Savings** Stop subscribing to IFR charts when you need only VFR charts or limit IFR chart coverage to your immediate area and the savings will be $30–$40 or more per chart cycle.

52 Checklist—personalized

✻ **Action** Make your own checklist on a personal computer, for use in the cockpit. In this day of desktop and laptop computers, word processor programs and graphics software, there is no reason why you should not make your own one-page checklist and have it laminated for $1–$2 (Fig. 3-11).

Figure 3-11

A homemade laminated checklist can be attached to the homemade clip.

Get the appropriate checklists out of your manual and lay them out on a single page with a suitable word processor. On the reverse of the normal checklist, print the emergency checklist. If you can't print back to back, photocopy the two checklists on one sheet back to back. Have the sheet laminated at your local stationery store (or they will be able to tell you where to go), and you are in business.

A word of warning: Do not amend the contents of the checklist in the aircraft manual. Reproduce every item in sequence faithfully.

$ **Savings** Given the going rate for commercially produced laminated checklists, you can expect to save at least $10.

53 Engine oil

* **Action** Purchase oil by the case rather than by individual 1-quart containers. You will save substantial amounts of money buying in bulk from any source. A friendly mechanic might help you tap into a wholesale source, in which case you will realize the greatest savings. Or a group of pilots might combine forces to get discounts from a wholesaler.

$ **Savings** At $2.70 per quart, 24 individual quarts of oil would cost $64.80. The same 24 quarts per case would cost $48 or less, a savings of $16.80 or more.

54 Flight bags

* **Action** Buy your flight bags in a discount luggage store or discount office supply store, rather than an aviation goods source. Large leather bags are also a tool of the trade for salesmen, lawyers, and accountants. High-volume stores catering to these buyers can afford to give the best discounts. It is not unusual to be able to buy a leather bag at a discount source for the price of a vinyl bag from an aviation goods store. Shop around.

$ **Savings** It would not be unusual to save $40 or more.

55 Flight planning log

✻ **Action** Make your own flight planning log on a personal computer using the table function of a word processor or spreadsheet. Reproduce the blank log on the photocopier. Easy and fun to make, custom-made flight planning logs can also be personalized to meet your specific needs (Fig. 3-12).

Figure 3-12

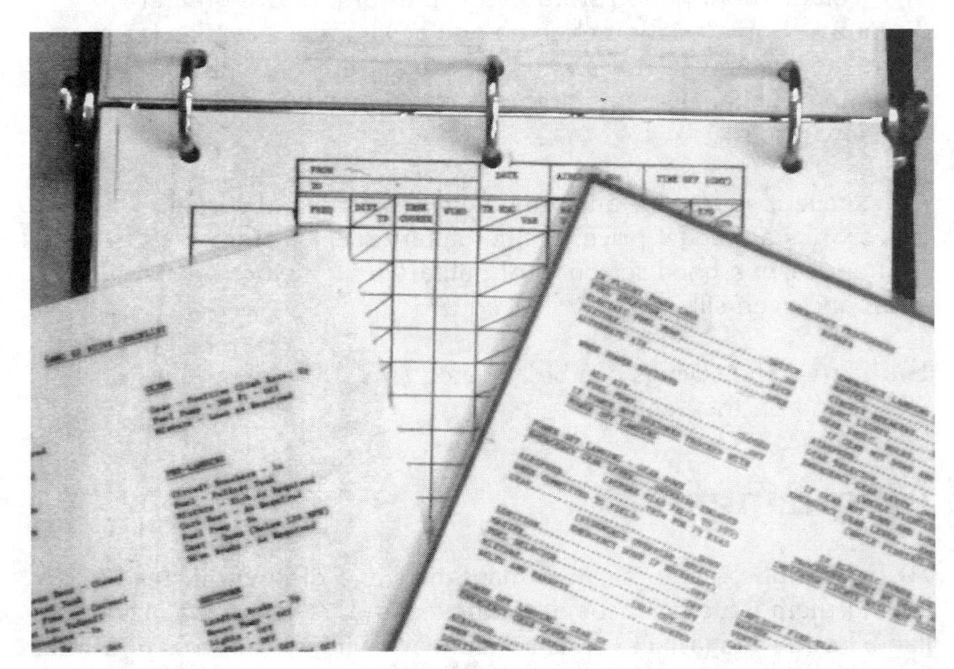

Flight planning logs are easy to make and can be readily customized.

A more sophisticated time-saving variation: Use a spreadsheet program to lay out the flight log, then enter in the required formulas to make all the calculations necessary to complete it. You can then enter the raw data and print out the completed log in no time before you leave for the airport. You can also store the regular routes that you fly and add only the wind information to update the flight plan for a specific trip.

$ **Savings** Dollar savings might be \$3–\$5 per set of 100, depending on your access to low-cost copying. (Time-savings in a quantifiable

dollar amount is obvious, but intangible until you determine what your time is worth and how much time might be saved by the customized spreadsheets.)

Clothing and accessories

* **Action** Buy certain items of flying apparel and accessories from a high-volume nonaviation store that can afford to give greater discounts. Some aviation clothing and accessories, such as leather flying jackets and aviator's watches, are extremely popular with the nonflying public. Discount clothing sources sell the merchandise at very low prices.

An excellent source for authentic flying apparel and certain accessories at budget prices is the military surplus store. Practically all of them carry a good selection of military-issue flying jackets, flight suits, and even silk scarves.

$ **Savings** The savings can be as much as $100 and up per item, depending on the item.

Magazines

* **Action** Subscribe to aviation magazines at a discount instead of buying them issue per issue at the newsstand. Aviation magazines, like all other magazines, offer substantial savings incentives to attract subscribers. Longer subscriptions yield greater savings per issue. Two discount subscriptions for one year might cost the same as buying only one of the magazines at the newsstand for a year.

$ **Savings** Two years of *Kitplanes* at the newsstand costs $84: subscription $44; *Private Pilot* for two years at the newsstand $71: subscription $38. You would save $40 and $38, respectively. (Prices for illustration only.)

58 Fuel purchase

✻ **Action** Always buy fuel from the least expensive source. Depending on airport size and regional competition, fuel prices can vary by as much as 25 cents per gallon. If your airplane is based at a large regional airport with all the bells and whistles, you might find that the little airstrip a stone's throw downwind is selling avgas for a lot less than your home field; many pilots make it a practice to land there and top off on the way home. Several flight planning software packages offer information on fuel prices.

Sometimes, if there is more than one fuel supplier at your home field, they might periodically engage in price competition from which you can benefit. Monitor those pumps, and don't hesitate to switch suppliers if the price is right.

$ **Savings** You would save perhaps $100–$250 per year (average four-seater flown 100 hours; 15–25 cents per gallon fuel savings). If the price difference at the pumps is 25 cents per gallon, you save a dollar every 4 gallons. So, if you burn 8 gallons per hour and fly 100 hours per year, you save $200. For that amount you can buy about another 100 gallons of fuel for a Skyhawk at $2 per gallon, which translates into another 12–13 hours of flying per year. Even at a fuel price savings of 12 cents per gallon, you will get an extra 5–6 hours per year in the Skyhawk. Is it worth it? Would you turn down 5 hours of free flying?

Fuel sampler

✻ **Action** If the fuel drain on your airplane is operated by pushing up on an external protrusion or rim, use a small jam or baby food jar as a fuel sampler (Fig. 3-13). (This suggestion will not work if a long and thin stem has to be pushed inside the drain opening to release fuel.)

$ **Savings** You save $4–$8, which is the typical price for a fuel sampler.

Figure 3-13

This mustard jar was free and is just as effective as any commercially available fuel sampler.

60 Gust locks

✱ **Action** There are several inexpensive alternatives to commercially made gust locks. Immobilize the control column with the seat belt; this is the age-old solution, and it is free.

Interconnect the right handle of the left yoke and the left handle of the right yoke with a bungee cord (Fig. 3-14). Make sure the tension is tight when touched lightly, but not so tight that the tension puts undue stress on the yokes. This is a very elegant solution. It costs only the price of the bungee cord.

Figure 3-14

A bungee cord makes a surprisingly effective and simple gust lock.

If you feel like building gust locks, that too is a simple task. A gust lock is easily made from ¼- or ½-inch plywood scraps, a thin bolt or eye bolt of appropriate length, some nuts and wide washers, and some felt or thin styrofoam.

Cut two squares out of the plywood (round off the edges) and glue a felt or styrofoam backing on one side of each piece. Drill a hole through the center of each piece. Make sure that the hole is just slightly larger than the bolt, so that when the bolt is stuck through it, the plywood can pivot somewhat to align with the wing/control surface.

Place a large washer on the bolt, then slide the two plywood pieces on it (felt/styrofoam side inward), followed by another washer and capped off with a self-locking nut or eyebolt fastener. Slide the bolt in the space between the control surface and the wing or empennage until the plywood pieces lie up snugly against the airframe surface, locking the controls in place (tighten the nut to suit). When the nut or fastener is properly tightened, the gust lock can be installed and removed with no further need for adjustment.

Warning: Be sure to put a highly visible streamer on the gust lock as a reminder to remove the lock before starting the engine.

$ **Savings** The ready-made option can cost as much as \$15. Depending on how good a scrounger you are, you should be able to make your own gust lock for, at the most, \$5. You save \$10 or more.

61 In-flight relief bottles

* **Action** A suitably clean plastic jar with a screw-on cap is an excellent budget in-flight relief bottle. A carefully chosen surplus jar with a proper shape and size for the task works just as well as containers made specifically for the purpose. Scrounge around in the kitchen. Do a "simulator session" with the jar at home before you try it in actual flight.

$ **Savings** The surplus jar is free for this use, so you save the cost of the alternative container made specifically for the purpose, as much as \$10, perhaps more.

62 Instrument covers for partial panel

* **Action** Make your own partial panel instrument covers. Cut out a suitably sized instrument cover from heavy cardboard, plastic sheet, or other opaque and lightweight material. Stick a Velcro dot on the perimeter. Stick another Velcro fastening dot on the instrument panel, just above the instrument you want covered. Attaching the cover is then a snap. Ideally, the instrument cover should be the same color as the instrument panel to blend in and not be a distraction.

$ **Savings** Scrounge for the cover material and hook-and-loop fasteners and you can make the covers for free. If you have to buy a sheet of cover material at an office supply store and a small bag of fastener dots at the hardware store (go to discount stores), you can still save a buck or two per instrument cover over the commercially available alternative. Typical savings would be \$3–\$5.

 Instrument markings—homemade

✻ **Action** If you fly an experimental aircraft, giving you great latitude in working on the airplane, mark ranges and specific values on instruments with colored, adhesive-backed plastic sheets or tape available in drafting supply stores.

Many instruments used in experimental aircraft come unmarked. You save money if the alternative is to exchange at considerable expense the unmarked instrument for one that is marked, or have a shop custom mark the instrument.

Carefully cut nice range lines (drawn first with a compass) with a sharp pair of scissors. Cut uniform hash marks with a sharp razor knife.

You can also use this technique to mark maneuvering speed range on any airspeed indicator as a personal reminder. Maneuvering speed range is never factory marked on airspeed indicators. (Remember that maneuvering speed is a range that varies directly with gross weight. Higher gross weight means a faster maneuvering speed. See your aircraft manual for specifics.)

$ **Savings** If the alternative is replacement with a commercially marked instrument, the savings can be $40 or more.

 Instrument panel covers—purchase

✻ **Action** For an instrument panel cover, buy a space blanket at a discount store that sells outdoor goods (Fig. 3-15). The primary purpose of a panel cover is to reduce heat by reflecting it. A space blanket reduces heat nicely for less than a cover sold specifically for the purpose.

$ **Savings** A space blanket on sale at a discount store is significantly less expensive than panel covers sold specifically for the purpose. You save up to $20.

Figure 3-15

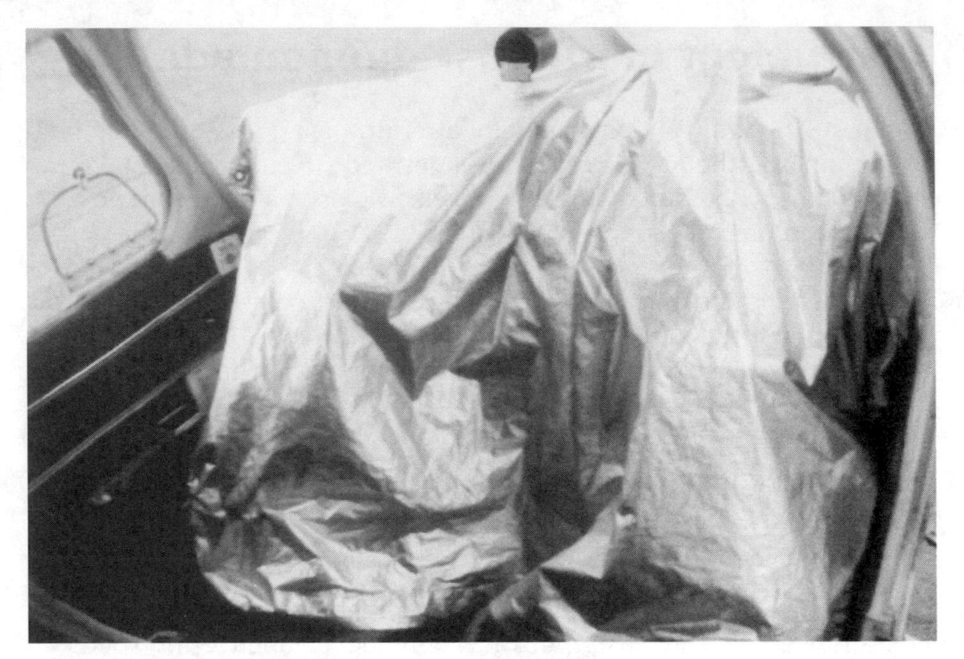

A discount space blanket is an effective budget instrument panel cover.

65 Intake plugs

* **Action** Buy a block of high-density hard styrofoam from a packaging materials supplier and carve intake plugs out of it. Make a paper template of the general frontal shape of the intake and with a bread knife carve an initially somewhat oversized plug (practice carving on a small piece before you tackle the big one that counts). Delicately keep shaving away to fit to size. Take great care not to slice away too much, ruining the plug. It should lodge in the intake so tightly that even fairly moderate tugging would not dislodge it.

The thicker a plug, the better it will stay in place. If space is available in the opening, get at least a 6-inch thick piece of foam.

To better preserve the new intake plugs, paint them with a heavy-duty heat-resistant paint; normal paint might bubble from residual heat if the plug is installed immediately after a long flight. One elegant solution is to sew colorful canvas slipcovers for the plugs.

$ **Savings** You will save approximately $20 compared to ready-made plugs.

66 Parts and accessories

✻ **Action** Purchase parts and accessories directly from discount supply houses. Your maintenance facility adds a hefty markup to parts and accessories it gets for you—as it well should, given the convenient service it is providing you. But if you are on a budget and work with a mechanic on that basis, you can come to an arrangement where you get certain parts directly from the least expensive sources.

Numerous accessories and parts are best bought in this manner: vacuum pumps, magnetos, fuel pumps, standby vacuum pumps, brake pads, brake discs, filters, strut overhaul kits, artificial horizons, heading indicators, tachometers, plus other flight and engine instruments. *Trade-A-Plane* is a good source of suppliers.

When available, buy overhauled parts and accessories from a reputable supplier or overhaul facility. Exchange the malfunctioning part or accessory for an overhauled unit. If time is not crucial, save even more by shipping the malfunctioning original to be overhauled, then have it reinstalled, rather than exchanged.

Refer to the next tip regarding core value and also refer to chapter 4 regarding major repair and overhaul.

A word of warning: It is imperative to get only parts for your certified airplane that conform to all FAA requirements pertaining to certified aircraft, and applicable to your specific aircraft type.

$ **Savings** Consider this. A vacuum pump was offered new by a maintenance facility for $345. The same pump was offered by a mail-order discount supplier for $305 new and $175 overhauled. Savings of 50 percent or more are not unusual when compared to prices at a local maintenance facility or parts supplier and the prices of new parts and accessories.

Parts and accessories—core value

✻ **Action** Do not automatically discard accessories you no longer need. Some have a substantial core value and are sought by overhaul shops that advertise in *Trade-A-Plane* for the accessory cores. Vacuum pump and alternator cores are in greatest demand.

Accessory overhaul facilities make a living refurbishing runout or malfunctioning accessories. When you need a replacement accessory, you are expected to send in the old one in exchange (or you get charged extra). The overhaul facilities also buy accessories outright. You might be throwing money away if you discard an old accessory. Check it out.

$ **Savings** Savings are measured by the core value paid for the old accessory. For example, vacuum pump core values at one time were \$40–\$90.

Parts and accessories—exchange

✻ **Action** If you need an overhaulable part or accessory and don't want to wait until it is overhauled, exchange it for an already overhauled unit instead of discarding your unit and buying a replacement outright. Many instruments, avionics, and accessories have respectable exchange values from which you might as well benefit. Check with suppliers and accessory or instrument overhaul shops.

$ **Savings** A person can save as much as \$50–\$100 per exchanged unit.

Pilot's clipboard

✻ **Action** Buy an inexpensive general purpose clipboard in a discount store instead of an expensive one supposedly made for pilots (Fig. 3-16). Slap a few strips of thin, flat weather stripping on the back to prevent it from slipping around.

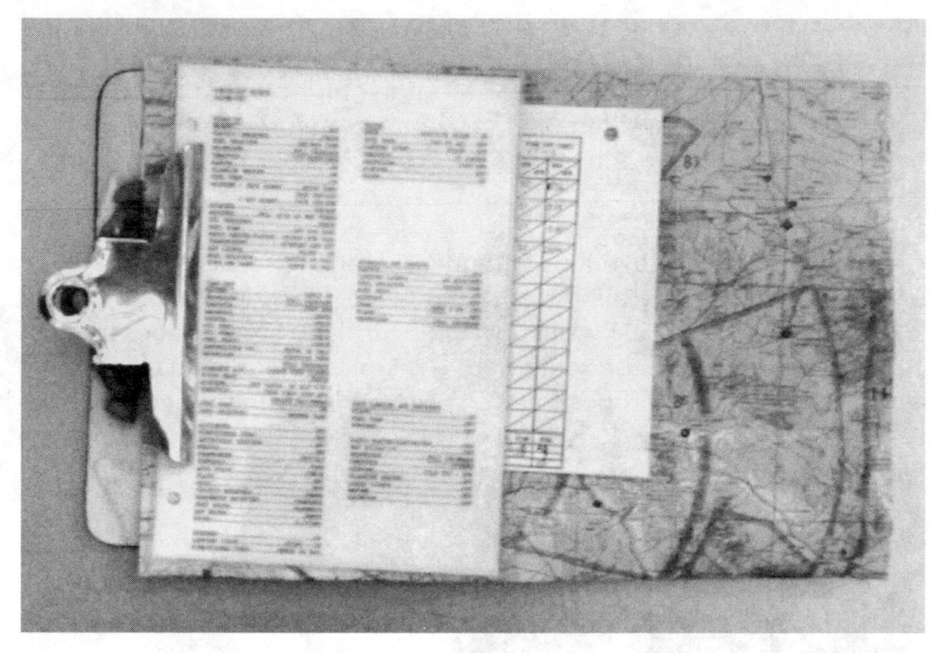

This drug-store clipboard has been serving in the cockpit for two decades.

A small general purpose clipboard with weather stripping on the bottom can be equipped with Velcro fasteners on elastic straps to function as a kneeboard. Buy the fasteners and elastic by the foot in a fabric and sewing materials store.

$ **Savings** You will save $10 and up. General purpose clipboards are available at discount stores for as little as $1.95. Simple clipboards marketed as being made especially for pilots can go for as much as $30.

Pitot tube covers

✻ **Action** Make your own pitot tube cover instead of buying one. There are several extremely easy options.

A word of warning: Make sure that the pitot tube cover you select permits pressure equalization to prevent damage to the pitot static system. If the fit is super tight, punch a pin hole in the cover to provide for pressure equalization.

Another word of warning: Brightly colored streamers attached to the cover are *extremely important*. Whatever option you choose to make the pitot tube cover, be sure to attach to it a long, brightly colored streamer as a safety measure to alert you to remove the cover before flight. A heavy-duty bright orange or red plastic ribbon works best.

Among the options, take a used tennis ball and carefully cut an opening into it with a utility razor knife (Fig. 3-17). Make the hole a size and shape that snugly slips over the airplane's pitot tube. In most cases the opening will be round, but the concept works equally well with the rectangular probes found mostly on low-wing Piper aircraft.

Figure 3-17

A tennis ball with a 1-inch slit makes an effective instant pitot tube cover. This one should have a streamer attached to remind the pilot to remove it.

Or buy a piece of tygon tubing of a circumference that snugly slips over the pitot tube. Cut to length, and seal one end with a dab of silicone, epoxy, or RTV usually found around the house or in your toolbox (don't spend any money on the sealant; scrounge around until you can borrow some if you don't already have some yourself). A ½-inch slice of a small wooden dowel of appropriate diameter glued in place also works well.

Cigar tubes, spark plug storage tubes, and spent shotgun shells might serve as an excellent pitot tube cover if the size matches. Please review the words of warning at the beginning of this tip.

A final possibility is a Naugahyde cover. If you want to get really fancy and have access to a heavy-duty sewing machine, you may make a cover from a piece of Naugahyde available in any store selling fabric or upholstery material. Make a pattern out of paper. Usually, two symmetrical pieces sewn together to form a "boot" will do the trick. The boot should slip on all the way to the point where the pitot tube meets the airframe.

Cut the Naugahyde to size (initially leaving a generous margin to make sewing easier) and sew the appropriate seams together. Trim the margins when sewing is complete. As an added touch, you can sew Velcro fastener strips inside the rim of the open end (this needs to be done before you sew the two pieces of the boot to each other). When you slip the boot in place and press the open end snugly together around the tube, the Velcro fastener will provide a tight seal ensuring maximum protection and making it more difficult for the cover to inadvertently slip off.

$ **Savings** Depending on your choice of alternative solutions you can expect savings of about $5–$15. Pitot tube covers can go for as much as $15; if a tennis ball that otherwise would be discarded does the job, that is $15 in your pocket available for avgas. Even if you have to buy some of the materials for the homemade cover, you can easily recognize a $5 cash saving. A Cessna 150 will fly for 30 minutes on $5 worth of gas; $15 will buy you gas for an hour and a half.

Preheater sharing

✻ **Action** Purchase a preheater together with fellow pilots and share it. Owning a preheater of your own is a waste of a good resource, given the little use it gets on one airplane. Preheating service from an FBO is certainly convenient, but hardly a budget buy. So why not band together to buy a preheater and spread the cost around? A bit of logistics is required for joint ownership of a preheater. It has to be placed in a secure area to which all users have access.

$ **Savings** If five pilots jointly purchase a $350 preheater, each pilot pays $70; each pilot saves $280.

Shipping costs

✻ **Action** Request the least expensive shipping option when you place a mail order. Shipping companies are cleaning up from the widespread but irrational phenomenon of "I want it now!" What's the rush? Plan ahead a bit and have the spare tire sent to you surface shipment for a fraction of the cost of next-day air.

Simply put, anticipate: "The annual is in a month, the left tire is getting bald, I better order a replacement today," instead of "I have to be in Tampa on Monday and the annual must be finished on Friday. Ship that tire overnight." Many suppliers include the cost of the least expensive form of surface shipment in the purchase price. Why pay extra?

$ **Savings** If surface shipping is included, you save the entire cost of rush shipment, typically $10–$20.

Side window replacement

✻ **Action** Replace the side windows with a stock plexiglass sheet of equivalent material specs and thickness, instead of buying hideously overpriced replacement windows from the manufacturer.

Side windows in most light aircraft are simple plexiglass sheets of a certain thickness, trimmed to size. Yet, if bought precut from the aircraft manufacturer, these windows are terribly expensive. Buy a plexiglass sheet that matches the original equipment specifications from a general plexiglass wholesaler for a fraction of the price, and use the old window as a template to cut the replacement window.

Note: Most side windows are made of flat plexiglass, others are slightly curved at installation to fit the contour of the fuselage. Certain windows in certain aircraft might have compound curvatures that cannot be replaced by a piece of flat plexiglass.

$ **Savings** Plexiglass sheets are cheap. Aircraft windows of equivalent material can easily cost more than a hundred dollars per window; therefore, the savings might be more than $100.

 Stopwatch timer

✻ **Action** Buy an inexpensive sports timer or stopwatch, suitable for attaching to the yoke with a Velcro fastener, instead of an expensive "aviation" stopwatch.

Inexpensive backlit, digital readout stopwatches are used in many athletic events and come in many varieties, even with minute countdown timers. They cost much less than the ones sold specifically for flying. Check out the features you want, and save. If you want elapsed time, be especially careful to get a stopwatch that measures hours. Many sports stopwatches don't go beyond minutes.

$ **Savings** You can save $20 or more per stopwatch.

 Suppliers—mail-order discounts

✻ **Action** Buy from high-volume mail-order suppliers to benefit from discounted prices. Comparison shop carefully. Mail-order suppliers also compete against each other and at any time one might be running a special.

Note that not all mail-order suppliers are necessarily discount sellers. Especially the glossier catalog companies are driven by the benefits to the customer of one-stop shopping and the widest selection of merchandise, rather than the best prices. There is also great variation in comparative prices from item to item. Again, shop around.

$ **Savings** Generally, you can save 10–15 percent or more compared to alternative prices.

 Suppliers—trade discounts

✻ **Action** Take advantage of trade discounts from suppliers for buying in bulk. A good example is the aforementioned purchase of oil by the case. Even greater savings might be realized, if, with a bit of organization, you can buy several cases at a time. Band together with

nine other owners and put in a single order for 10 cases. Let it be known that you expect the deepest discount possible.

A buyers' association of aircraft owners might also be able to get trade discounts on other supplies, such as brake fluid, lubricating oil, aircraft degreasing products, hand cleaners, charts (everyone in the buyer's association needs a new sectional on the same day), windshield cleaner, and aircraft washing detergents and wax.

A group of owners loosely organized into an association might be able to negotiate a discount on fuel prices at the local airport if they guarantee a certain amount of fuel purchases per month.

Also be alert, as an individual owner, for the common practice of trade discounts offered on fuel to locally based aircraft.

$ **Savings** Expect 10–30 percent savings or more over normal retail prices.

77 Tiedown kit

✳ **Action** Make a portable tiedown kit. The components of a portable tiedown kit are available in every discount store with a hardware section. You need three anchors, most commonly used to temporarily anchor Fido's leash to some spot, and 30 feet of nylon rope cut into three equal pieces. Tie each rope to an anchor, and you are in business.

$ **Savings** You should save about \$10–\$15 compared to commercially available portable tiedown sets. To realize these savings, it is important to source the anchors and rope in a big discount store offering hardware, rather than the local mom and pop hardware store.

78 Tires

✳ **Action** Purchase retreaded tires instead of new ones when the treads on the tires you have begin to fade. The expected life span of a retread is equivalent to what you should get out of a new tire, and it is a lot less

expensive. If you are concerned about reliability, consider that the airlines and the military make widespread use of retreaded tires.

Opting for the retreads also helps the environment. Retreading uses fewer resources and pollutes less than the manufacture of new tires.

$ **Savings** Specific savings depend on the number and type of tires you need, perhaps $50–150. On a full set of tires for a typical four-seater you can realize savings of up to $150. On average, retreads go for about half the price of new tires.

Tools—buying

✻ **Action** Most tools used to work on aircraft, especially in routine owner-performed maintenance, are generic. Buy them from a discount tool supplier rather than an aviation supplier. Also watch for the tremendous deals the general discount sellers regularly advertise.

Larger sales volumes allow the generic discounter to price tools more competitively than an aviation supplier. For this reason, aviation suppliers don't offer too many general-use tools, but don't be tempted by the tools that they do offer, without first checking nonaviation discount sources.

You can also save, even at a general discounter, by opting for lesser brand name tools, if you don't use them every day. A less expensive Phillips screwdriver might wear out sooner than the top of the line brand name, but if you use it only to unbutton the airplane once a year at annual time instead of doing commercial maintenance work six days a week, chances are it will last as long as you have the airplane.

$ **Savings** You should save 30–40 percent or more, compared to alternative sources. A wire snipper offered for $13 by an aviation supplier cost $6 on sale at the generic discounter; safety wire pliers offered for $41 cost $20; a small flashlight offered for $18 cost $9.

Tools—sharing

✻ **Action** Band together with other pilots who like to work on their airplanes and share a comprehensive collection of tools. This arrangement might require the joint purchase of tools and is most practical for pilots who do serious work on their airplanes, such as experimental aircraft enthusiasts. A variation is to jointly buy a major item such as a drill press or band saw.

You can't be a crabby, possessive nitpicker about tools. Practical arrangements have to be made to ensure equal access for all users. The hangar of a group member is the most ideal solution.

$ **Savings** The personal savings will range from $30–$40 all the way to hundreds. If three pilots jointly buy a $200 band saw, each pilot saves $67.

Upholstery

✻ **Action** When it comes time to install a new interior in your favorite aerial mount, don't head for the local upholstery shop if you are on a budget. Purchase an upholstery kit and install it yourself. One supplier in particular specializes in upholstery kits, precut and preshaped for a wide range of aircraft, and ready for simple installation out of the box.

As with any budget solution, do your homework carefully to avoid disappointment. Write to the suppliers for information and material samples. Ask them where the finished product can be viewed and take a good look before you commit. Ask around at your airfield to gather information on wear and tear and tips on installation from pilots who have used the product.

If you absolutely do not want to have anything to do with installing upholstery, you can still save a substantial amount by buying a kit and having an upholsterer install it instead of having custom upholstery made.

A word of warning: If you do your own upholstery work, it is imperative to use only FAA-approved fabrics and materials. Failure to do so is not only a violation of law, but can also result in a serious safety hazard (fire) and could compromise your insurance coverage under certain conditions.

$ **Savings** Savings on an upholstery kit for an average four-seater, rather than a custom upholstery installation, can range from around $750 if you use the services of an upholsterer to install the kit, to as much as $2,000 if you do all the work yourself. Savings for larger and smaller aircraft will vary proportionally.

82 Wheel chocks

✳ **Action** Wheel chocks are extremely simple to make from scrounged wood or metal, instead of buying them.

For wooden chocks, start with a 6-inch-long piece of 2 × 2 wood. Saw the piece of wood by cutting diagonally from corner to corner the full length, which makes two triangular pieces lengthwise, and you have a set of wheel chocks.

For aluminum chocks, start with a foot-long 2 × 2 piece of aluminum angle iron. Cut it in the middle, which makes two 6-inch pieces, and you have a set of wheel chocks (Fig. 3-18).

To keep each pair of chocks together, it is worth connecting them with a piece of rope. Drill holes at one end of each aluminum chock, burr it out to smoothen the edges and attach the two chocks with a rope. Screw an eye bolt screw into one end of each wooden chock and attach the two chocks with a rope. Leave enough slack in the rope to put the chocks underneath the tires on your airplane and also give you something to easily grab hold of to remove them.

$ **Savings** If you can scrounge around for scraps and leftovers from other projects you can build a set or two of chocks for free, saving as much as $20 per set (for one wheel). Even if you have to buy the material, you will come out ahead of the ready-made alternative.

Figure 3-18

These wheel chocks were made by sawing a piece of aluminum angle iron in half.

83 Window shields

* **Action** Protect your instrument panel by making window heat shields from space blankets bought in discount outdoor equipment stores. Placing heat shields on all the windows to keep heat out of the cockpit is a more effective way of protecting the instruments than draping the instruments with a cover (Fig. 3-19).

Figure 3-19

Discount space blankets can be used to make window heat shields.

A couple of inexpensive space blankets have enough material to make window shields at a lower cost than a kit that is sold exclusively for the purpose. Another material that has become extremely popular lately, is a bubble-cell plastic sheet sandwiched between two thin sheets of aluminum. The bubble-cell material is available by the yard from mail-order suppliers.

Make paper templates first and use them to cut the shields to size. Place tracing paper over the outside of the windows to make the templates, which is much much easier than doing it from the inside. (After going to that much trouble, save the templates for any future needs.) For best results, fold over the edges of the shields and sew the border for reinforcement. Attach the shields to the windows with Velcro fastener dots.

$ **Savings** A set of ready-made heat shields for a light aircraft can cost as much as $80. For half that, or less, you can opt for the homemade alternative, saving up to as much as $40.

4

Aircraft use

84 Aircraft controls—rigging

* **Action** Make sure that the aircraft controls are in proper rig to ensure best aerodynamic performance. If the control surfaces are out of rig, the airplane will not fly straight and level when the controls are in neutral. Some control input is then required (achievable by trim, if so equipped).

The displaced controls generate additional drag in straight and level flight. The airplane's performance deteriorates because it needs higher power settings and uses more gas to match the properly rigged airplane's performance. At similar power settings, the airplane flies at a slower speed.

To give you an idea of the difference in performance, consider this actual example: An out-of-rig Arrow achieved 125 knots at 75 percent cruise; properly rigged, at the same power setting, altitude, and temperature, the cruise speed increased to 135 knots, a gain of 8 percent.

$ **Savings** A 5 percent savings on fuel expenses is a typical example when an airplane is properly rigged. For a ballpark estimate of possible savings, let's use the Arrow in the example. A 600-mile cross-country flight in the Arrow at 135 knots takes 4.4 hours, requiring 44 gallons of fuel; if the controls are out of rig, resulting in a cruise speed of 125 knots at the same power setting, the same trip takes 4.8 hours and requires 48 gallons of fuel. At a fuel price of $2 a gallon, the fuel bills are $88 vs. $96. The savings of 4 gallons amounts to $8 in your pocket at the end of the flight.

85 Autogas

* **Action** If an STC is available for your aircraft that permits the use of autogas, obtain the STC and use autogas in accordance with the STC rather than buying the more expensive aviation fuel. When fuel prices shot up in the 1970s, the price increases set into motion an energetic movement to gain approval for the use of less expensive autogas in suitable aircraft engines. As a result, many older types of aircraft that are equipped with low-compression engines, received STCs authorizing the use of autogas.

Using autogas in an aircraft approved for it continues to be one of the most popular savings measures of pilots on a budget. The annual savings are considerable.

Words of warning: Autogas should be used strictly in accordance with the terms of the STCs under which use is authorized. There are several important safety issues to consider regarding the use of autogas.

First, the engine as well as a production aircraft has to be STC'd for the use of autogas. Never use autogas in an engine and aircraft that are not STC'd.

The second safety issue: Pilots of experimental aircraft equipped with engines that are STC'd for autogas need to evaluate the effect of autogas on parts of the aircraft other than the engine because the FAA does not issue STCs for experimental aircraft. There is

considerable debate on the long-term effect of autogas on certain fuel tanks that are made of composite materials. Equally credible arguments can be made to suggest that autogas is not a problem, or, in the long run, it might affect the material composition of composite fuel tanks and become contaminated. Empirical data to prove either opinion is scarce.

If you are considering the use of autogas in the appropriately STC'd engine of your experimental airplane, absolutely check with the airframe designer first.

A third safety issue: Quality control of gasoline that was mixed for use in an automobile, rather than an airplane engine, might be a problem. Autogas specifications allow a wider range in the proportion of ingredients, and autogas quality control is less stringent than aviation fuel quality control. If the autogas meets its claimed specifications, you are all right; if it doesn't, you could experience combustion problems. There have also been some well-documented cases of gas stations fraudulently selling cheap, low-octane gasoline as the high-grade stuff. An FBO is going to be concerned about contaminants in aviation fuel; the corner filling station might be less concerned about contaminants in gasoline. Let the buyer beware.

$ **Savings** As much as 70 cents per gallon is a typical savings, which can be several hundred dollars per year. In an airplane that consumes 5 gallons an hour, the 70-cent-per-gallon fuel savings amounts to a savings of $3.50 per hour. Fly 100 hours a year and save $350. If your airplane consumes 10 gallons per hour, a 70-cent-per-gallon savings translates to $700 for 100 hours of flying per year.

Bartering for aircraft use

✱ **Action** Bartering is a good way to save because two people can barter at cost, exclusive of the profit margins built into normal pricing. Mechanics and flight instructors stand the best chance of bartering their services for flying time in a private owner's airplane. Accountants, bookkeepers, and lawyers might also find good barter opportunities.

An attractive form of barter to broaden flying experience is for owners of two vastly different aircraft types to trade flying time. Good examples are twin time for aerobatic time, or warbird time for floatplane time.

$ **Savings** You might save as much as 20 percent off the cost of for-profit services. By exchanging services at cost, both parties get a good deal. The trick is to fairly establish "cost."

87 Best rate of climb

✳ **Action** Climb to cruising altitude at the best rate of climb speed. (To review ground school: The best rate of climb speed is the speed at which the aircraft reaches cruising altitude in the shortest amount of time.)

Savings is realized by cutting down the time spent on a segment of the flight that consumes fuel at a high rate. The sooner you reach cruising altitude, the sooner you can set up best economy cruise to benefit from the associated fuel savings.

There is usually only one recommended climb power setting for most light aircraft: maximum continuous power. Options in climb technique are provided by varying climb speed: best angle, best rate, recommended cruise climb. The same power setting yields the same high fuel flow at any speed; therefore, to use minimum fuel during climb, spend the least amount of time possible climbing. You can achieve this objective by climbing at the best rate of climb speed.

The usual trade offs for climbing at best rate of climb are increased engine temperatures and reduced forward visibility. Some aircraft are severely affected, others not at all. The bottom line: If in your aircraft it is inadvisable to continuously climb at best rate of climb speed for whatever reason, climb at an acceptable speed closest to best rate of climb speed.

$ **Savings** Several dollars per climb can be saved. If your average rate of climb at the best climb speed is 800 fpm, you would reach 8,000 from sea level in 10 minutes (8,000 ÷ 800). At a "cruise

climb" (shallower climb, same power setting) of 500 fpm, you reach 8,000 feet in 16 minutes (8,000 ÷ 500). If you climb at best rate of climb, you spend 6 minutes (one-tenth of an hour) less at maximum continuous power. At a fuel flow of 12 gph, 1.2 gallons of fuel is saved, or $2.40 assuming a fuel price of $2 per gallon. If you climb to 8,000 feet from sea level 100 times a year, the saving is $240.

Braking technique

✻ **Action** Go as easy on the brakes as you can to keep wear and tear to a minimum, extending the life of the brake pads. Brakes are tempting things to stomp on. It is a lot easier to manhandle the airplane and correct the error of your ways by standing on the brakes, than to pay attention to good technique and hardly touch the brakes. Yet, with a little practice good braking technique can become second nature and will greatly extend brake pad life. The suggestions that follow will get you started in the right direction.

Don't taxi at a power setting so high that you have to "ride" the brakes to keep taxi speed sufficiently slow. It is so easy to retard the throttle to a setting that gives a comfortable taxi speed, yet so many pilots abuse the brakes instead.

Anticipate the need to slow down while taxiing. Retard the throttle to idle early, and lightly brush the brakes to come to a stop when the airplane has slowed considerably, rather than motor along and stand on the brakes seconds before you ram the stationary airplane ahead of you.

Don't use the brakes to turn if the airplane is equipped with a steerable nose or tailwheel. It is so easy to jab the brakes to get a turn going. Avoid the temptation. Steer only with the steering mechanism (usually the rudder pedals) firmly and with sufficient anticipation.

Don't smoke the brakes on landing to impress an imagined audience by your ability to make aircraft carrier landings. The only comment you will get from those whose opinion you should value will be "Why did you put a year's wear and tear on the brakes in one landing?"

Don't consistently land so long that you have to brake "firmly" to come to a stop. The best technique is to touch down at the proper touchdown speed a little beyond the numbers (having had a good margin of safety over approach obstacles), leave all drag-creating devices fully deployed on roll out, and let the airplane roll to the end of the runway. Except at really short fields, the airplane will practically roll to a stop on its own accord.

$ **Savings** You can save as much as $100 per set of brake pads. If you can double the life of brake pads by learning proper braking technique, you cut the need for replacing brake pads in half. Given parts and labor costs, your savings will be significant.

Cross-country—aircrew discount card

✻ **Action** Obtain an aircrew card available to any pilot, and receive hefty discounts on hotels and motels when flying cross-country. Aircrew cards are available for a modest annual fee from pilot organizations such as AOPA. The cards are honored at thousands of participating hotels that offer room-rate discounts to pilots.

$ **Savings** You might save as much as 40 percent off regular room rates. These savings are not to be ignored, especially in big cities worldwide where regular hotel prices are outrageously inflated. A 35-percent discount on a $180 per night hotel room in Washington, D.C., is a savings of $63.

90 Cross-country—airport courtesy cars

✻ **Action** Use the FBO's free courtesy car to run into town for a bite to eat, instead of taking an expensive cab. At many airports, including small ones where taxi service might be erratic, the local FBO has a courtesy car available to visiting pilots for short-term use. At the smaller airports, the car is often "good transportation," one that has seen better days but is a perfectly adequate runabout.

Often this service is not widely publicized, but a polite query could bring forth car keys instantly. AOPA's *Aviation USA*, *Flight Guide*, and *JeppGuide* contain information on the availability of courtesy cars at airports.

Don't expect to be lent a car for a 3-day tour of the local national park system. Courtesy cars are for quick trips to go to local restaurants, or to run an errand or two. Many FBOs, especially at the more remote airports, will also let visiting pilots take the courtesy car overnight to the local hotel on a first-come first-served basis.

$ **Savings** Typical savings is $30. Cab rides of $15 each way between the airport and town are quite common these days, even in small-town America. That's $30 in your pocket if you get a courtesy car.

91 Cross-country—airport guides

✻ **Action** Purchase one of the leading airport information guides, such as AOPA's *Aviation USA*, *Flight Guide*, or the *JeppGuide* (Fig. 4-1). Information in these reasonably priced guides is invaluable when planning cross-country flights economically. In addition to specific airport information including landing and parking fees, they all contain motel, restaurant, transportation information, reservation numbers, and much more.

Figure 4-1

Two information-packed airport guides.

You save by being able to plan your flight to benefit from accommodations and services best suited to your budget, rather than showing up and finding out what's available when you get there. Traveler beware: Make telephone calls a week or so prior to departure to verify that the information is current.

$ **Savings** Researching the information guides might save hundreds of dollars per trip.

92 Cross-country—airport/hotel courtesy vans

* **Action** Make extensive use of free courtesy vans provided by airport area hotels, instead of taking expensive cabs. Most of us are familiar with the courtesy van arrangement around major airports. This service is available at even the smallest, most remote airports these days if there is a well-known hotel chain in town or if there is heated competition between the local establishments.

Always be sure to ask for a courtesy pick up when you call the hotel. AOPA's *Aviation USA*, *Flight Guide*, and *JeppGuide* contain information on the availability of hotel courtesy vans at airports.

$ **Savings** A courtesy van can translate into $30–$40 per hotel stay. A $15 cab fare each way between the airport and nearby hotels is not unusual.

93 Cross-country—avgas prices

* **Action** Plan refueling stops at airports with the lowest avgas prices. This is a very popular savings technique and hinges on your ability to find out avgas prices in advance along your route of flight. Fortunately there are several sources of such information, and with the ever increasing popularity of on-line computer services, more sources are likely to spring up in the near future.

Some of the flight planning software programs offer fuel price information, such as Flitesoft's Bargain Fuel Locator. A service called Fillup Flyer Fuel Finder is available on-line to computer equipped subscribers via modem, or by mail to those who don't have computer access. The nice thing about Fillup Flyer is that it is becoming available at FBOs, where you can check ahead for fuel prices as you prepare for the next day's flight plan.

To be economical, your fuel savings have to be sufficiently high to exceed the cost of getting the information, but usually any cross-country trip of several hundred miles will do the trick.

$ **Savings** Buy 50 gallons of fuel at a 50-cents-per-gallon savings—as a result of your research—and you are ahead $25 minus the cost of obtaining the information. Including subscription rates to obtain the information, the savings on the first 50 gallons might get you to break even, but even a couple of cross-country trips a year will make these services worthwhile.

94 Cross-country—camping vs. hotel stays

✻ **Action** If you are the adventurous type, camp out next to your airplane (or sleep in it if it is big enough) instead of staying at hotels/motels. Camping is more easily done at the smaller, out-of-the-way airports in semirural, rural, and wilderness areas. Fortunately, the United States is so sparsely populated that even near the country's biggest cities you are rarely more than 30 air miles or so from an airport where you can probably camp. Most airports that will let you camp out rarely charge extra fees beyond the overnight parking fee.

Always get permission to camp. Pick small airstrips and call ahead. Many will let you taxi off to a grassy area and set up overnight camp. Some might even have shower facilities.

A good budget strategy is to intermingle camping days with motel stays as circumstances permit. Pack a couple of foldable bikes and you are really in business.

A word of warning: Be conscious of personal safety considerations, and don't hesitate to ask about security concerns. Personal safety should not be a concern at the majority of airports where camping is allowed, but you can never be too careful in our crazy world.

$ **Savings** Savings per night are equivalent to what you would spend for a hotel room: $40–$250 per night. On a two-week cross-country vacation, assuming an average nightly hotel bill of $60, two nights in a hotel and the other nights camping, your savings are $720.

95 Cross-country—landing fees

* **Action** Land at airports that have no landing fees or charge modest fees in comparison to neighboring alternatives. Landing fees vary widely. As a rule of thumb they are low or nonexistent at the small, uncontrolled fields and can be hefty at the megajetports or main tourist attractions. AOPA's *Aviation USA*, *Flight Guide*, and *JeppGuide* all contain landing fee information at all airports nationwide. Plan accordingly.

$ **Savings** You would typically save $2–$10 per landing.

96 Cross-country—overnight parking fees

* **Action** Land at airports that have no overnight parking fees or charge modest fees in comparison to neighboring alternatives. Some larger airports can charge fairly stiff overnight fees, so it is worth finding out in advance. AOPA's *Aviation USA*, *Flight Guide*, and *JeppGuide* contain information on overnight parking fees at all airports nationwide.

The published information, however, might not always be accurate, and at times might only indicate that there is a fee, without providing an amount. Could it be that if you have to ask, you can't afford it? Call ahead to avoid surprises and flight plan accordingly.

$ **Savings** Nightly savings are typically $5–$15.

97 Cross-country—picnics vs. airport food

✻ **Action** Pack a picnic instead of buying expensive airport food. Picnics are easy to pack, can be as simple or as elaborate as you can afford, and even on the elaborate side the cost is competitive with airport restaurant prices. **A word of warning:** Ensure proper preparation and storage of all foods to prevent food spoilage and the possibility of food poisoning. Safety of the trip, especially a long trip of several days, might be seriously compromised by illness of the pilot or a passenger.

$ **Savings** A picnic for two would realize a $5–$10 savings. Two tuna sandwiches and two glasses of milk cost $12.50 or more at many local fly-in restaurants. The same meal packed at home costs $5, a savings of $7.50. Or, to look at it another way, for $12.50 and a little work in the kitchen you can put together a lot fancier picnic than two tuna sandwiches and two glasses of milk. The picnic plan yields a lot more food for your money.

98 Cross-country—portable bicycles

✻ **Action** Buy two portable bicycles (or one if you fly alone) to save on taxi rides to interesting spots at your destination. This idea is a long-term savings scheme because you have to make a considerable investment in quality bicycles. When you recoup the cost, the taxi-ride money that would be spent stays in your pocket.

The bicycles will also make your flying more interesting. Knowing that you have the bicycles will make you more willing to fly places where you have to get around. The bicycles will also put you and your companions in superb shape.

Good examples of ideal bicycle destinations are Block Island off the Connecticut coast and Nantucket off Cape Cod. Many pilots fly in explicitly to rent bicycles and end up taking $15 taxi rides each way between the airport and the bicycle rental places.

$ **Savings** Typically you would save a minimum $10 when the initial investment is repaid and you do not have to pay for a taxi. Assume that the average round-trip taxi ride from the airport and back again costs $20. If you bought two bicycles for a total cost of $500, 25 round trips to town pay for the bicycles. An avid touring pilot easily makes that many trips in a year, beyond which each bicycle ride means money in the pocket. In some instances it might take two or even three years to pay for the bicycles, but they are usable for many, many years after that.

99 Cruising altitude

✳ **Action** Cruise at the most fuel-efficient altitude for your airplane (to the extent that it is practical and safe, given your mission). Generally, the higher you go, the lower the fuel consumption to attain the same performance. In the thin, high-altitude air, less fuel is required to attain the correct fuel-air mixture. Consult your aircraft operating manual.

Table 4-1 shows the fuel consumption in gallons per hour at different altitudes and power settings for the Cessna Skyhawk on a standard day. (Consult your aircraft manual for specific requirements.)

Table 4-1

Cessna 172, fuel consumption (gal/hr)

Altitude	2,200 RPM	2,400 RPM
2,000 ft	6.3	8.0
6,000 ft	5.9	7.2
10,000 ft	5.6	6.9

The higher the fuel consumption of your aircraft in comparison to other aircraft types, the more the difference in fuel consumption at various altitudes, and the better the hourly savings.

$ **Savings** The Skyhawk's fuel consumption table shows that at 2,200 RPM it uses 0.6 gph less at 10,000 feet than at 2,000 feet, a per-hour savings of $1 (assuming avgas at $2 per hour). In 100 hours of cross-country cruise per year, your savings would be $100.

At a faster cruise setting of 2,400 RPM, the Skyhawk uses 1.1 gph less at 10,000 feet than at 2,000 ft, a per-hour savings of about $2.20 (assuming avgas at $2 per hour). In 100 hours of cross-country cruise per year your savings are $220.

Now look at the third scenario: flying at 2,000 ft at a power setting of 2,400 RPM vs. flying at 10,000 ft at 2,200 RPM. The per-hour difference in fuel consumption is 2.4 gallons, a savings of $4.80. In 100 hours of cross-country cruise the savings is $480.

Delivery flights

* **Action** Volunteer for delivery flights for expenses only. There is a great need to move aircraft from one part of the country to another. Airplane brokers, FBOs, and private individuals all need to have an airplane moved from time to time, and might not have a pilot available on the spot to do it. Often, these airplanes are flown by pilots who volunteer their time for the flying experience, and are compensated only for their expenses.

A variation on the solo ferry flight is accompanying another pilot, such as a new owner taking delivery of the airplane and feeling uncomfortable in it solo on a long flight home.

The trick is to find out about these opportunities and to be available to take advantage of them. How successful you are depends in large part on how good you are at getting the word out about your availability, and having the flying experience required.

A private pilot's certificate is sufficient for volunteer ferrying services because you are not being compensated. You need to have solid experience in the type of aircraft you propose to ferry, but an instrument rating is not crucial. Remember, there are plenty of VFR-only aircraft that need ferrying.

Here is an actual example: A private pilot, comfortable in motorgliders, but not excessively experienced, was asked by the dealer to fly one of the prettiest composite motorgliders from the East Coast to the EAA convention at Oshkosh and back. The pilot got 22 hours of some of the finest flying he ever did, just because he was competent in type and available.

$ **Savings** Savings are in the hundreds of dollars, if not more. Measure savings in this case by comparing the ferry flight time to paying your expenses out of your pocket for the same trip. When you add up fuel costs, hotels, food, and aircraft rental or the ownership costs allocated to the period of the flight, your savings are in the hundreds of dollars, or more.

101 Descent technique—avoid shock cooling

* **Action** Descend at a sufficiently high power setting to avoid shock cooling of the engine and a premature overhaul. It is tempting to just retard the throttle to idle and point the nose at the ground when you want to come down, but it is also asking for expensive trouble. The cylinders are vulnerable to rapid temperature changes. Metal expands and contracts as the temperature goes up and down. Too sudden a temperature change and the metal might not be able to keep up. It cracks. The rapid cooling with the power off is likely to cause cracked cylinders, especially if the procedure is repeated over and over again. So, avoid shock cooling. Follow the aircraft operating manual's recommended descent technique.

A valuable aid when monitoring descent technique is a digital cylinder head temperature gauge. A rule of thumb for the average air-cooled, horizontally-opposed light aircraft engine is a temperature drop of no more than 1° per every 3 seconds. Count "one thousand, two thousand, three thousand...", observe the temperature drop degree by degree, and take immediate corrective action if necessary. The 1° per 3 seconds is only a rule of thumb. Consult the aircraft or engine manufacturer for specific recommendations.

$ **Savings** Typical specific savings are difficult to cite; you are likely to save the cost of unnecessarily overhauling one or more cylinders before TBO, at an expense of hundreds of dollars, if not more. Simple repair of a cracked cylinder is expensive; the top overhaul is worse and a major overhaul would still be necessary when the time comes. Avoid shock cooling and save probably hundreds of dollars.

102 EGT and CHT gauges

✳ **Action** Equip your airplane with exhaust gas temperature (EGT) and cylinder head temperature (CHT) gauges to maximize efficient engine operations. Only with these gauges can you properly lean without running the risk of cooking the cylinders. Leaning without these gauges is like shooting in the dark, a rough guess.

Equipping each cylinder with separate EGT and CHT gauges is best, but the single gauge installed on the hottest cylinder (which is not really the hottest under all circumstances) is an effective and less costly alternative (Fig. 4-2). Digital gauges are the most accurate; a good budget compromise is a combined digital EGT/CHT gauge installed on the hottest cylinder.

Figure 4-2

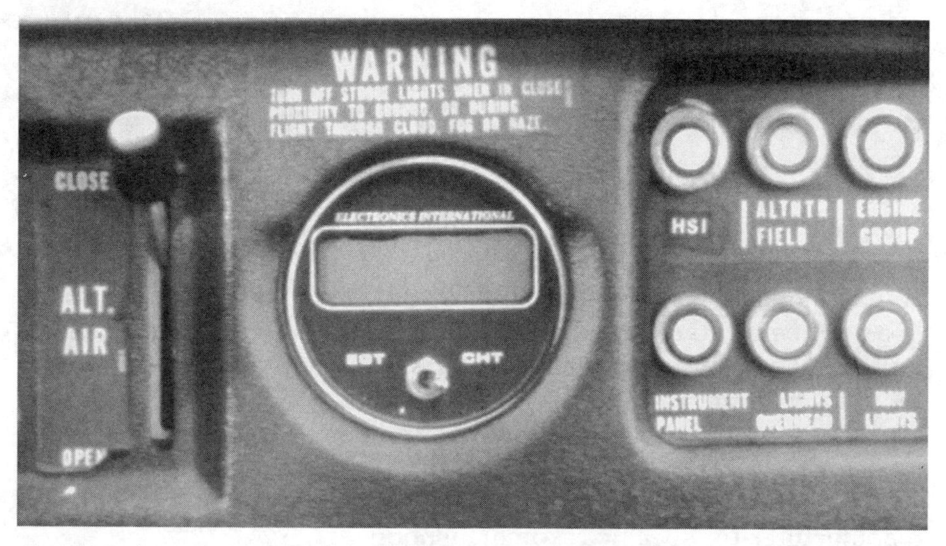

Combined single probe EGT/CHT gauge is a good budget compromise.

Independent gauges on each cylinder provide an early warning of looming trouble. Any deviation from expected temperature patterns can be investigated and might lead to the discovery of a problem well before it becomes an in-flight emergency. An alarmingly rapid temperature change in-flight might be noticed early enough to make a precautionary landing before an unmanageable engine problem develops.

$ **Savings** You could save tens of dollars or more in fuel savings per year due to accurate leaning with the gauges. Perhaps you would save hundreds of dollars by postponing maintenance because you monitor temperature trends and keep them within certain operational parameters.

Engine—ground operations

⁂ **Action** Operate the engine on the ground as briefly as possible. Never operate the engine on the ground at high-power settings for prolonged periods of time. Most aircraft engines are air-cooled. They suffer without the airflow that streams over them at flying speed.

$ **Savings** Tens of dollars or more would be saved due to prolonged engine life. Specific savings are difficult to measure, but are significant in giving you more use out of the engine.

104 Engine—leaning on the ground

⁂ **Action** Save fuel and reduce the chance of plug fouling by leaning the engine during taxi. Fuel flow has a cooling effect, so be careful to not overheat the engine by leaning the mixture while you taxi. Keep an eye on the EGT and CHT gauges (some aircraft are more affected than others) to keep temperatures in check during taxi. Lean conservatively if the aircraft does not have EGT and CHT gauges.

A safety concern is forgetting to enrich the mixture prior to takeoff. A good habit is to leave your hand on the mixture knob during taxi and upon reaching the runway set the mixture as appropriate for takeoff, before you remove your hand.

$ **Savings** You could save \$20–\$30 or more on spark plug maintenance between inspections. Servicing fouled plugs is expensive. Reduced fuel consumption will also help you realize some savings.

105 Engine—starting

✳ **Action** Two easy-to-perform precautionary actions decrease wear and tear and contribute to prolonging engine life. **Warning:** Because you will be turning the propeller by hand, make sure the mags are off before touching the propeller!

Turn the propeller by hand one compression cycle for each cylinder in the engine to coat the cylinder walls with oil (four cycles for four cylinders, six for six, etc.). You will feel tension as you turn the propeller blade, followed by a release of tension. When the tension is released, one compression cycle is completed. Then, when you start up you minimize metal grinding on raw metal.

Just prior to getting in to start the engine, turn the propeller backwards until you hear the impulse coupler click. This will set up the engine starter to turn the prop immediately, saving wear and tear on the starter and conserving battery power.

$ **Savings** A dollar savings is difficult to quantify, but probably significant due to prolonged engine and starter life.

Engine—warm-up

✳ **Action** Warm up the engine properly before runup. Make sure all gauges are in the green. In most cases engine warm-up is not a problem because the weather is not exceptionally cold and the engine warms up sufficiently during the long taxi to the runup pad. But if the airplane is tied down very close to the end of the runway and you are anxious to get underway in very cold weather, it might be tempting to perform the runup before the engine gauges all reach the bottom of the green. This is hell on the engine and should be avoided.

$ **Savings** A dollar savings is difficult to quantify, but could be significant due to prolonged engine life.

107 Engine cooling modifications

⁕ **Action** Install any STC'd engine cooling modifications that are available for your aircraft to reduce engine wear and tear. Few things are worse for an engine than frying it. The most common cause of engine overheating is poor air circulation. Engines that are prone to overheating experience greater metal fatigue than properly cooled engines and generally require an overhaul well before TBO.

Most of the aircraft with poorly designed cooling systems date back to the 1940s and 50s. Two good examples are the Globe Swift (Fig. 4-3) and the Navion. Alternative cowlings with improved engine cooling have been developed for both aircraft and are highly recommended.

Figure 4-3

This Globe Swift's highly modified engine cooling system is a blessing for the engine.

$ **Savings** Savings are best measured in terms of getting more use out of the engine, more value for your money: perhaps hundreds of dollars. The cost of the modification also has to be taken into account, but it is not as severe as might first be imagined because it adds significantly to the resale value of the aircraft. At any rate, the savings realized by getting an extra 700–800 hours out of an engine by improving its cooling, is in the hundreds of dollars.

108 Engine use—minimum flight hours per month

✱ **Action** To get the most time out of your engine before overhaul, use it frequently, and put a monthly minimum number of flying hours on it. An engine that isn't used for weeks at a time suffers almost as much as one that is abused. During periods of inactivity the moisture in the cylinders isn't dried out and can cause rust, the various lines and hoses dry out and might crack, and the battery might run down. Also, the pistons might not seat properly, leading to low compression, and you might end up needing an overhaul sooner than you expect.

A rule of thumb for light aircraft is that an engine should fly at least about 15 hours a month to stay in good shape. Check with the manufacturer. If you aren't flying 15 hours a month, increasing the number of flying hours is hardly a form of saving, although it might cost less to fly more often than to overhaul the engine well before TBO.

To get the monthly time up where it should be, perhaps you need another partner or two. If you don't increase flying time, you stand a good chance of paying for at least a top overhaul before recommended TBO.

A word of warning: Running the engine on the ground for prolonged periods is detrimental. Do not run the engine on the ground to increase time on it. Many people don't want to go flying in the dead of winter and get the idea of running the engine on the ground to give it a little workout. This is damaging to the engine and should not be practiced. Go flying instead.

$ **Savings** The saving is realized in the cost of the extra maintenance or premature overhaul that you manage to fend off by increasing flying time. It can be hundreds of dollars.

109 Expenses—sharing with passengers

✻ **Action** Share the expenses of a flight with your passengers to cut down on your per-hour flying costs. According to the FARs, sharing the cost of a flight with the passengers is not flying for hire; therefore, it can be done by private pilots.

The savings for the pilot can be quite high depending on how much flying is shared, and are a nice subsidy of the pilot's hourly flying costs.

$ **Savings** You could save tens of dollars per flight. Suppose the fuel bill for a 5-hour weekend round-trip with three passengers is $100. If the cost is shared, at the end of the flight, your passengers give you $75.

Expensing aircraft use—business deductions

✻ **Action** If you are using an aircraft in your business, deduct the expenses of owning and operating or renting. Expensing the business use of an airplane is perfectly legitimate if done according to the rules, but it is also greatly abused, inviting attention from the IRS.

To qualify for tax deductions there are two general rules of thumb. First, the business on behalf of which the aircraft is being used has to be a legitimate, properly established and documented venture with a demonstrable chance of profitability. The business can experience losses year after year, but the IRS must be convinced by the nature of its activities that it will turn a profit eventually. The business may be full-time or part-time and it may be your primary or secondary source of income, as long as it has a shot at profitability.

The second rule of thumb is that the specific activity for which the deduction is being claimed has to be an integral part of operating the business rather than, say, a vacation masquerading as a business trip.

The first step to deducting flying expenses is consulting your tax accountant. Describe what you want to do, find out what you can legitimately claim given your specific situation, and set up the record keeping properly.

$ **Savings** You could save hundreds of dollars or more. For instance, you are a builder with sites in five neighboring states and you spend $20,000 a year flying between them in your Mooney. If you are in the 30 percent tax bracket, and deduct the $20,000, you save 30 percent of $20,000 in income tax expense, which is $6,000.

111 Expensing aircraft use—hobby tax deductions

✻ **Action** If you don't qualify for general business deductions you might still be able to deduct some of your flying expenses under the hobby tax rules. If you run a business as a hobby, for instance your primary objective is to engage in the activity rather than to make money from it, you may offset any gross income derived from the hobby by an equivalent amount of the expenses it took to generate that income.

Now, in English: If you took aerial pictures of the neighbors' houses primarily to amuse yourself rather than to make money and the neighbors paid you $2,000 for the pictures, you may deduct $2,000 of the expenses incurred in taking the pictures. It might have cost more to take the pictures, but under the hobby tax laws you can deduct only as much in expenses as you made in gross income. You can offset income but you can't deduct a loss. The IRS places time limits on how long a hobby can be unprofitable.

The first step in deducting your flying expenses under the hobby tax laws is consulting your tax accountant. Describe what you want to do, find out if it is a "hobby" as defined by the tax laws, find out what you can legitimately claim given your specific situation, and set up the recordkeeping properly.

$ **Savings** Hundreds of dollars or more could be saved. If you can deduct in flying expenses the $2,000 you were paid for taking aerial portraits of the neighbors' houses, and you are in the 30 percent tax bracket, your savings in income tax expense are 30 percent of $2,000, or $600.

112 Flap seals

⁕ **Action** Install flap seals to increase the aerodynamic efficiency of the airframe. Save on fuel by using less fuel to attain the same performance. Savings will come in small increments from this proposition. The increase in performance is rarely more than 5 percent, which usually translates into a lower percentage in fuel savings (Fig. 4-4).

Figure 4-4

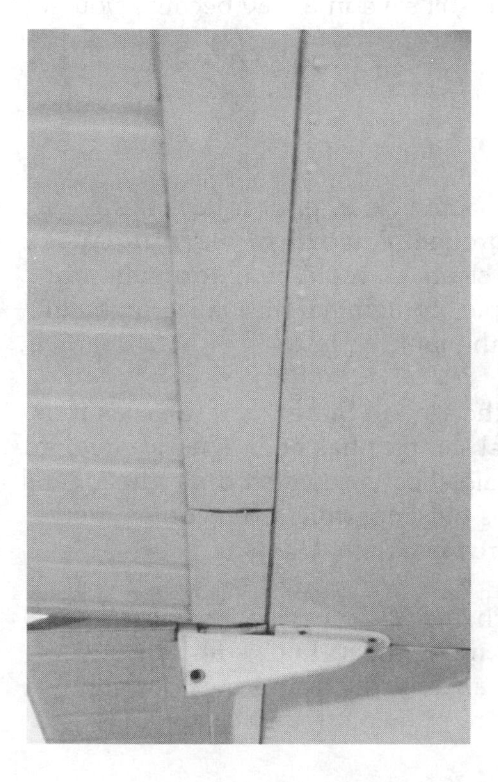

Flap seals add a few knots at the same power setting, or yield the same speed at a lower power setting which saves fuel.

The cost of the seals also has to be considered: there is the up-front investment but it is at least partially recouped by increasing the resale value of the airplane. In spite of the up-front investment and the incremental performance increases, the flap seals will still probably put a few dollars in your pocket at the end of the day. Calculate the costs and the savings for your own specific situation.

$ **Savings** Tens of dollars per year could be saved, possibly more savings when the investment in the flap seals is amortized. If flap seals reduce the fuel bills for a Cherokee by 3 percent and the total fuel bill for a year is $4,000, the total savings is $120 per year. If the seals cost $200 and half of that will be recouped in increased resale value, they pay for themselves in the first year, after which the fuel savings go straight to the proverbial bottom line, your wallet.

If you choose to fly faster on the same amount of fuel, rather than fly at the same speed on less fuel, you still save in a way because you get more "time" value for your money.

113 Fuel sampling

✳ **Action** When you sample fuel and it is clear, pour it back in the gas tank instead of pouring it on the ground. **A word of warning:** Make sure the fuel sampler is spotlessly clean. It would be unfortunate, to say the least, to draw a clear sample, contaminate it inadvertently in the sampler, and dump it back in the tank.

In Europe and elsewhere outside the United States, where avgas is as much as $6 a gallon, recycling fuel samples has been a fairly standard practice for years. It has recently also begun to catch on in the United States. One hi-tech sampler strains out impurities from contaminated fuel when the sampled fuel is poured back into the tank.

There is also an environmental benefit. All fuel spills, however small an amount, are hazardous to the environment. Fuel is terribly toxic. Even if it is poured on the tarmac and quickly evaporates, it might cause more harm than we are aware of by getting into our lungs.

$ **Savings** Recycling the fuel sample should save at least a few dollars a year. Sure it takes some time to reach a whole gallon of fuel samples, and a gallon is only about $2 or so. But nobody can deny that pouring any clear fuel on the ground is literally throwing money away, even if it is only a few cents at a time.

114 Hoerner tips

✻ **Action** Exchange the wingtips of your aircraft for aerodynamically more efficient Hoerner tips. You will get a few extra knots out of the airplane or you can elect to fly at the same speed at a marginally lower power setting, thereby saving on fuel. (Other benefits of Hoerner tips are marginally increased lift and better slow speed handling).

Hoerner tips are expensive and even with the increased resale value it would take forever to save their cost in fuel; therefore, from a budget point of view, the financially most efficient time to install Hoerner tips is when the old, factory- installed tips crack beyond repair (as they eventually do on most light aircraft), and must be replaced.

$ **Savings** You might save a few dollars a year.

115 Manage an aircraft

✻ **Action** Find well-to-do but busy aircraft owners and arrange to take care of their aircraft in exchange for flying time. This arrangement might best work with a group of partners who need a partnership coordinator. Tasks can include scheduling, billing, bill paying, aircraft cleaning, washing, and waxing, and arrangement and supervision of maintenance. (Note: Make sure that the full partners become responsible for collections if any partner fails to pay properly.) If you are an instructor or mechanic and can offer these services, your chances of landing a management arrangement might be especially good.

Another way to increase your chances is to put special effort in selling the concept of regular cleaning (it could even be a weekly progressive program) as a means of keeping the aircraft spotless and preserving its value.

A good opportunity to build flying time under a management agreement is delivering and picking up an airplane for maintenance at another field. A large number of owners have maintenance arrangements with what they feel is the best facility within 100–150 miles of their home field. Getting the airplane to and from the mechanic's shop is always a production because of scheduling conflicts, which becomes a good opportunity for an aircraft manager to get some flying time.

Aircraft management tasks at the light aircraft level are usually a part-time endeavor and might not justify unlimited access to flying time. There are several compensation options. First, accurately place a dollar value on your services and agree upon a block of time representing fair compensation. A second alternative is to arrange for a heavily subsidized hourly rental rate in exchange for management services.

A third variation is to, in effect, become a junior partner by contributing your management services instead of equity. You pay your share of the day-to-day fixed and operating expenses of the partnership but own no share of the airplane. Options are to put a limit on flying time and accepting a lower reservation priority than the equity partners.

If you have the time, you can offer your services to several groups and get block time commitments in a variety of aircraft. The variations seem endless. Be creative and politely persistent.

$ **Savings** If you are lucky enough to land an aircraft management arrangement, the total savings—in comparison to what it would cost you to buy the same type of flying—is substantial, easily hundreds of dollars per year.

116 Maximize flight experience per hour

* **Action** Pack as much training and as many refresher exercises as possible into each flight. It isn't total flight time that counts, but what you do during the time you spend in the air. An hour of touch-and-goes is of much greater value than a two-hour straight-and-level cross-country flight along a coastline.

The military routinely packs its sorties with a multitude of training tasks to get the most out of its flying dollars, and so should you. Here are some suggestions.

Establish a list of related training tasks before each flight, carefully designed to keep you current and sharp in all aspects of your flying (Fig. 4-5).

Figure 4-5

The confusing engine control console on this old aircraft is a small price to pay for budget performance but the console does require careful transition training.

Be creative with cross-country routes. Fly segments at different altitudes; use different navigation techniques per segment; design specific navigation exercises, such as VOR intercepts and establishing your position via VOR bearings; try to hit your flight plan as closely as possible.

Go out once a month for touch-and-goes; practice a variety of takeoffs and landings: short field, soft field, crosswind, and so on.

Plan your flights to include regular stops at challenging fields. Regularly incorporate emergency procedure practice into your flights.

Enhance your routine IFR flights in IMC with training flights that include full procedures, NDB approaches, partial panel exercises, and the like. Make sure that you have the required safety pilot on board; perhaps you could swap places after a coffee break.

$ **Savings** You might save as much as several hundred dollars per flight time required to be safe and pass checkrides or flight reviews. Staying in top form saves money two ways. First, it cuts down on expensive flying time required to be as safe as you can be; second, it cuts down on expensive training flights when you decide to get additional ratings and pilot certificates.

117 Oil coolers—install the option

✻ **Action** If your aircraft is not equipped with an oil cooler and there is an approved installation, install one. Cooking the oil drastically reduces its lubricating effectiveness. If the oil temperature is constantly on the high side, your engine is likely to need a top overhaul well before recommended TBO.

A good example is the Grumman TR-2. Without an oil cooler, which is the standard factory configuration of the airplane, TR-2s usually need a top overhaul at 1,400–1,500 hours. When an oil cooler is installed at the outset or early in the engine's life, the recommended 2,000–hour TBO is very likely.

$ **Savings** Savings of hundreds of dollars or more are realized by not having to have an expensive top overhaul performed prior to recommended TBO.

118 Oil quantity—topping off

✻ **Action** When you top off the oil in your aircraft, time it to take a quart so that the last half-quart of capacity remains unfilled. If, for example, the oil sump holds 8 quarts, add 1 quart at 6.5 quarts.

This practice saves money in many aircraft because the first half-quart is quickly blown through the breather, and is, in essence, wasted. (Monitor the oil consumption carefully when filled as the manufacturer recommends to see if the airplane blows out the first half-quart.)

$ **Savings** Tens of dollars or more per year might be saved with careful oil quantity management. If your aircraft uses a quart every 8 hours and you top off to the sump limit, you might be losing half a quart through the breather tube every 8 hours. At $2.60 a quart that is $1.30 down the drain every 8 hours. If you fly 250 hours a year, you are losing $40 unnecessarily.

119 Power setting—best economy

✻ **Action** One of the most effective ways to stretch your dollars is to cruise at best economy setting whenever you can. There are two power settings that should be used in most light aircraft, best economy and best power. Both power settings are achieved by proper leaning technique and can be applied at any percentage of power.

Best economy is the setting at which the aircraft achieves the desired power using the least amount of fuel. Best power is the setting at which the aircraft achieves the fastest speed. The choice between best economy and best power depends on your mission objective.

The difference in fuel burn between best power and best economy settings can amount to as much as 2 gallons and that translates into a lot of money; typically more, in fact, than the savings per gallon on competitive fuel prices.

Table 4-2 shows the fuel consumption figures at best economy and best power settings for a Piper Arrow III at a variety of percent power settings. (Consult your aircraft manual for specific needs.)

Table 4-2 **Piper Arrow III, fuel consumption (gal/hr)**

Percent power	Best economy	Best power
75%	10.2	11.6
65%	9.2	10.4
55%	8.2	9.1

As you can see, best economy saves more than a gallon of fuel per hour at the same power setting. If you fly at the lowest power setting at best economy, rather than at the highest power setting at best power, your hourly fuel savings are 3.4 gallons of fuel per hour.

In the Arrow, best economy is achieved by leaning to peak exhaust gas temperature and best power is achieved by leaning to 100° rich of peak EGT. For the appropriate technique to set best economy and best power on your aircraft, see the operating manual, or the engine operating manual.

$ **Savings** You could save $200–$400 or more (based upon an average four-seater flown 100 hours per year, saving 1–2 gallons per hour). By cruising at a best economy setting instead of best power, you can save as much as $4 per hour if fuel costs $2 per gallon and you reduced consumption from 10 gallons per hour to 8 gallons per hour. If you fly 100 hours a year, you will realize a 20 percent savings on your fuel bill. Or, if you choose to spend your savings on more flying, you will increase your flying time for the year by 20 percent at no extra cost.

Power setting—constant-speed propeller

✻ **Action** There are two ways to achieve low power settings in aircraft equipped with constant-speed propellers: first, high RPM and low manifold pressure; second, low RPM and high manifold pressure. When you face this choice, choose the second option, low RPM and high manifold pressure.

The lower RPM prolongs engine life by subjecting it to less wear and tear. There is also a benefit to this technique during descent. It keeps the engine warmer and reduces the possibility of shock cooling.

Consult the aircraft operating manual for the power setting options available on your airplane and act accordingly.

$ **Savings** Prolonged engine life will foster substantial savings. It is difficult to quantify precisely the potential savings resulting from this power setting technique, but the prolongation of engine life yields more value for the money invested in the engine.

Power setting—effect on tachometer

✻ **Action** If you have the option of renting two aircraft at the same hourly rate, but the time is measured in one aircraft by tach time and another by Hobbs meter time, always opt for the airplane available on tach time. The same "hour" on tach time is longer in real time than an "hour" of Hobbs time.

Tach time is measured by engine RPM. At low power settings the time drum (the little numbers on the tachometer) rotates more slowly, registering less time. Hobbs meter time is driven by engine oil pressure, which is generally constant, regardless of power setting.

Measured time differences depend on a variety of circumstances, but as a rule of thumb, for every hour of Hobbs time you'll get 1.2 hours of tach time.

Practically all commercial operators measure time by the Hobbs meter; however, flying clubs and private owners often use tach time.

$ **Savings** Generally, you should expect to save about 10–20 percent per hour at the same rental rate. At an hourly rate of $70, measuring it by tach time instead of Hobbs time means $7–$14 per hour in your pocket.

122 Power setting—percent power

✻ **Action** Always set a low percentage power setting to conserve fuel. Regardless of whether or not you are leaning to best power or best economy (as noted elsewhere in power-setting tips), you save fuel by flying at a low power setting.

Table 4-3 is an example of fuel consumption by the Piper Arrow at different power-setting percentages. (Consult your aircraft manual for specific needs.)

Table 4-3 **Piper Arrow III, fuel consumption (gal/hr)**

Percent power	Best economy	Best power
75%	10.2	11.6
65%	9.2	10.4
55%	8.2	9.1

As you can see, at 55 percent power, compared to 75 percent power, the airplane consumes 2 gallons per hour less when leaned to best economy and 2.5 gallons per hour less when leaned to best power.

$ **Savings** On an hourly basis, you could expect to save $2–$6 per hour. If your hourly savings are $4 and you fly 150 hours per year, your annual savings are $600.

Power setting—proper leaning

✲ **Action** Always follow proper leaning procedures to get the most efficiency out of your airplane and to prolong engine life. Improper leaning can result in either a mixture that is too rich or too lean. Each condition has its shortcomings.

If the mixture is too rich, you are unnecessarily burning too much fuel—you might as well be burning money; if the mixture is too lean, the engine temperature might become excessive—causing damage to the engine over the long run.

Different phases of a flight and different aircraft require different mixture settings. Follow the aircraft operating manual for proper leaning procedures.

The only really effective way to lean precisely and monitor the mixture is by using exhaust gas temperature (EGT) and cylinder head temperature (CHT) gauges. The most useful EGT/CHT gauge is the type that measures EGT and CHT for each cylinder, but one gauge that monitors only the hottest cylinder is also very useful.

$ **Savings** Fuel savings might amount to $1–$2 per hour. Hundreds of dollars could be saved during the life of the engine due to averted maintenance.

Preheating

✲ **Action** Always preheat the engine when the temperature approaches freezing. A sure way to ask for premature cylinder wear and tear is to grind together the brittle, cold-soaked metal of the pistons and cylinder walls.

Every aircraft manufacturer publishes a recommendation for engine preheating at or below a certain cold temperature. Adhere to the recommendation. Or establish a warmer temperature to initiate preheating, which can only help the engine.

$ **Savings** You could save several hundred dollars by averting premature maintenance.

125 TBO—exceeding a recommended TBO

✻ **Action** Time between overhauls is strictly a recommended limitation. If the engine shows no signs of trouble, continue to operate it beyond recommended TBO, but monitor it with exceptional care.

Many aircraft routinely go several hundred hours beyond recommended TBO. If the engine has been well cared for, the oil regularly changed, and the proper operating procedures observed, the chances are good of getting some extra time out of it.

Consult with your mechanic around recommended TBO, get a thorough professional evaluation of the engine—especially an oil analysis, and if it passes, keep on flying.

$ **Savings** Several hundred dollars in deferred maintenance can be saved. If an overhaul costs $15,000 and you flew for 2,400 hours on an engine with a 2,000-hour TBO, your engine overhaul cost per hour works out to $6.25 vs. $7.50 per hour.

126 Throttle operation—proper technique

✻ **Action** Open and close the throttle smoothly and gradually. Quite a few pilots are tempted to bang the throttle back and forth with great gusto. There must be something addictive about the resulting burst of power, or they have been watching too many movie pilots power up to save the world on the late show.

Yet pounding the throttle forward is like trying to swallow your dinner all at once instead of bite by bite. The suddenly increased fuel flow is traumatic for the engine. Snatching the throttle all the way back to idle with a flick of the wrist is just as detrimental.

Rapid changes in throttle position greatly increase the wear and tear on the engine and can lead to expensive maintenance before its time.

$ **Savings** Forestalling the need for premature maintenance could save several hundred dollars in the long run.

127 Tires

✻ **Action** Treat the airplane's tires well. Always have them properly inflated, minimize their abuse by practicing proper landing and braking technique. A pilot can do several things to maximize the longevity of the airplane's tires.

Proper inflation is the most effective method for prolonging tire life. Underinflated tires wear much more rapidly because tire's profile is different from the design profile and experiences greater stress from the forces acting on it. Improper inflation can reduce tire life by as much as 50 percent. Buy a tire pressure gauge (ideally with a flexible hose for easier handling and a dial indicator for easier reading) and learn how to use it.

Hard landings cause tires to wear prematurely because of the excessive friction generated between the stationary tires and the ground. Certain transport category and military aircraft are equipped with a mechanism that initiates the tire rotation prior to touchdown. This would be overkill on light aircraft, but we can at least soften the blow with consistently soft landings.

Hard braking also causes excessive friction between the tires and the ground and leads to premature wear. Stay well ahead of the airplane and brake lightly.

Here is a neat trick to reduce wear on the tires of retractable-gear aircraft. After takeoff and just prior to retraction, apply the brakes briefly to stop the tires from rotating. Otherwise they will hit a stop in the wheel well while rotating, which causes excessive wear.

$ **Savings** Typically you could save 25–50 percent in tire expenses. If a tire costs $40 and installing it costs another $40, by doubling the tire's life you are saving $80 per tire life cycle. (You would need two tire changes instead of one for the same amount of flying, if you mistreated the tires).

128 Volunteer flying

✳ **Action** Take advantage of volunteer flying opportunities to fly your airplane for the reimbursement of your expenses. There are a variety of volunteer flying opportunities (Fig. 4-6). Among them are flying out-patients requiring specialized treatment to distant medical appointments, flying search and rescue missions and cadet orientation flights for the Civil Air Patrol (you have to join the CAP), volunteering for fire patrol flights during exceptionally hazardous, dry periods, and volunteering to fly observers on wildlife surveys.

Figure 4-6

A volunteer tow pilot gets to fly an L-19 Bird Dog for free at the local soaring club.

Contact the following organizations for more information: flying physician associations for information on noncritical medical flights; local post of the Civil Air Patrol; state forestry commission; state wildlife commission; and nonprofit wildlife organizations such as the Sierra Club (ask them for information on wildlife organizations in your area).

$ **Savings** Hundreds of dollars in paid-for flying expenses. If you fly 10 hours on a volunteer wildlife patrol and get $35 per hour for your expenses, you get $350 worth of flying for free.

Wheel fairings

✻ **Action** If your aircraft is equipped with wheel fairings, leave them on the landing gear (Fig. 4-7). Wheel fairings improve the airplane's aerodynamic efficiency. They can increase cruise speed by 3–5 percent, resulting in more speed for the same amount of fuel burned (more value for your money), or less fuel required to attain the same speed (a cash savings in fuel expense).

$ **Savings** Typically expect to save a couple of dollars per flight in reduced fuel requirements or increased performance for the same amount of fuel.

Figure 4-7

Wheel fairings reduce drag. The increased efficiency saves fuel at the same airspeeds.

5

Aircraft maintenance

Part 43 of the Federal Aviation Regulations authorizes the holder of a pilot's license to perform certain preventative maintenance on the aircraft owned or operated by the pilot, provided that the aircraft is not used for Part 121, 127, 129, or 135 operations. Each tip describing maintenance items authorized by FAR 43 quotes the relevant authorization up front.

It is the sole responsibility of anyone performing any maintenance on any aircraft to be fully informed of and adhere to current maintenance regulations.

In addition to Part 43 preventative maintenance, owners can take a variety of additional measures to save on maintenance, also covered in this chapter.

130 Airframe refinishing

✻ **Action** FAR Part 43 permits "refinishing decorative coating of fuselage, balloon baskets, wings, tail group surfaces (excluding balanced control surfaces), fairings, cowling, landing gear, cabin, or cockpit interior when removal or disassembly of any primary structure or operating system is not required."

This FAR allows you to do a fair amount of repainting as long as it is preventative, you stay away from balanced control surfaces and refrain from taking anything apart. It is good news not only when it comes time to repaint the airframe, but also when you need to do touch-up work, repaint local repairs, or want to add an accent to the existing paint job.

Be sure to tackle painting only within the limits of your abilities, regardless of what the FAR allows you to do. Do not hesitate to head for the professional paint shop for the big jobs, but work out an arrangement allowing you to participate.

Obtain the advice of your mechanic for specific touch-up procedures for the kind of paint on your airplane. Generally, the steps are strip the old paint, clean the area, apply the primer, and apply several coats of paint.

Warning: Given the potentially hazardous consequences of inadvertently and inappropriately painting balanced control surfaces, always seek the OK of a qualified mechanic for any painting before you do it.

$ **Savings** You could save $25 and up into the hundreds. The savings potential is measured in labor expenses saved, not the amount of time it takes you to do the job, but the amount it would have taken the shop to do it for you. A small local touchup, for which you would have been charged an hour of labor at, say, $40, is $40 in your pocket instead. If you do a good chunk of the stripping and most of the masking on a major paint job, tasks that take several hours, you save a lot more.

131 Airframe repair

✻ **Action** FAR Part 43 permits "making small simple repairs to fairings, nonstructural cover plates, cowlings, and small patches and reinforcements not changing the contour so as to interfere with proper airflow."

This gives the pilot permission to repair nicks, minor dings, and small holes in the airframe. Typical examples are the replacement of cracked inspection plates, slightly damaged nonstructural fiberglass fairings, and small holes in the airframe skin.

The repair of metal skin usually consists of a patch riveted over the crack or hole. Cracks have to be stop-drilled before being patched. Generally, rivets must be at least three times the skin thickness in diameter and the distance from the damage to each rivet hole must be at least two rivet diameters.

Check with a qualified mechanic to make sure you are allowed to do the repair you propose and to confirm the specific standards to which you should adhere.

$ **Savings** Airframe repairs could save you $25 and up. Even the smallest nonstructural repair can take at least an hour or two of a mechanic's time. Some small jobs are quite simple, but can be time-consuming. Two hours saved at a $40 per hour shop rate is $80 in your pocket.

132 Airframe—washing

✻ **Action** Wash the airplane regularly. The more frequently you wash the airplane, the longer that the finish and the airframe will last. Grime, acid rain, other forms of pollution, and salt in the air do a number on the paint job, and eventually on the airframe. Use only detergents recommended in your aircraft manual and by the manufacturers of the finish on your airplane.

$ **Savings** Washing saves hundreds of dollars. Suppose a paint job costs $3,000. If you let it go, you might have to repaint the airplane in 5 years. In that case, the paint job cost you $600 per year. This means you have to save at a rate of $600 per year to have money for the paint job.

If you take care of it, you might be able to stretch the paint job for 8 years. In that case, the paint job cost you $375 per year; this is the amount you would have to save per year to pay for the paint job when the time comes; a savings of $225 per year ($600–$375) during the first 5 years of the paint job.

In the end you have to come up with the same amount of money, but if you make the paint job last longer you get more use out of the original paint job and you have a longer time to come up with the funds for the new paint job. You save because the cost of the paint job is a smaller portion of your income over 8 years compared to 5 years.

Airframe—waxing

❊ **Action** Wax your airplane once a year. By waxing your airplane periodically, in addition to washing it, you prolong the paint job's life even further, getting even more use out of the finish. Waxing is highly recommended as a finish preservation measure, especially if your airplane is tied down outside.

You save because you have a longer time to come up with the money to repaint the airplane; in comparison to a shorter time, every year the amount you have to set aside is a smaller percentage of your income.

Use only the wax that is recommended by the aircraft manual or the maker of the finish on your airplane. It even pays to have a professional wax job done, if you don't have the time.

$ **Savings** You could save hundreds of dollars for the same reasons explained in the previous tip, aircraft washing.

134 Annual—owner participation

✻ **Action** Do all the monotonous and routine work yourself on the annual. Do more if you can, to the extent allowed by the FARs. Most mechanics charge a flat fee for the inspection portion of the annual (with an extra charge for any necessary parts and labor). This generally hefty inspection fee includes the labor of unbuttoning the airframe to permit inspection, as well as reassembly following inspection, tasks that account for most of the labor inclusive of the flat fee. If you perform these tasks yourself, you might even be able to cut the inspection fee in half (Fig. 5-1).

Typical tasks owners can perform on the annual are: opening up all inspection holes; removing upper and lower cowling; removing tailcone; removing all fairings; draining oil, removing oil filter, removing oil and fuel screens for inspection and cleaning, removing seats, cockpit panels, and floorboards for access to items to be inspected; lubricating hinges and bearings, repacking wheel bearings; reassembling all these mentioned items following the A&P's inspection.

As you can see, there are at least a few hours of work in these tasks, all of which have to be performed just as part of the routine inspection. If any work needs to be done and the mechanic is willing, you can participate in that too, to the extent permitted by the FARs, in a mechanic's apprentice role, very closely supervised by your mechanic.

$ **Savings** You could save \$200 and up. At a shop rate of \$40, performing work that your mechanic would have taken 5 hours to do (it might take you 10), you are saving \$200.

135 Antimisfueling devices

✻ **Action** FAR Part 43 permits "the installation of antimisfueling devices to reduce the diameter of fuel tank filler openings provided the specific device has been made a part of the aircraft type certificate data by the aircraft manufacturer, the aircraft manufacturer

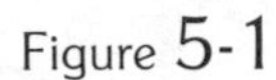
Figure 5-1

Supervised owner participation can significantly reduce the cost of an annual.

has provided FAA-approved instructions for installation of the specific device, and installation does not involve the disassembly of the existing tank filler opening."

Perform this task yourself and you are bound to save an hour or two in labor. This installation ensures that the aircraft does not get filled up with jet fuel when you need avgas, or vice versa. The device is most useful for turboprops, which are commonly confused with reciprocating-engine aircraft because of the propellers.

$ **Savings** The savings would be $40 and up. Beyond that, it is hard to put a price on your peace of mind.

136 Battery charger—solar

⁕ **Action** Purchase a solar battery charger to keep the battery trickle charging between flights. This proposition results in savings for owners who don't fly their airplanes frequently and have to deal with a weak battery when they attempt to start the engine after a prolonged break between flights. A battery that loses its charge and is then recharged rapidly, as it is with an alternator, develops discharge problems and has to eventually be replaced before its time.

A trickle charger keeps the battery charged during long periods of inactivity, eliminating the need for extra battery servicing and the premature need for a new battery. It also saves wear and tear that the starter motor suffers as you grind away on a weak battery.

$ **Savings** You could annually save from $20 or $30 to nearly $200. A solar charger costs $30. Savings are difficult to quantify, but if the charger eliminates the need for periodic extra servicing at, say, $10, and prevents the need for a new battery that costs $180, it already saves more than it cost.

137 Battery maintenance

⁕ **Action** FAR Part 43 permits "replacing and servicing batteries." This is a small, simple task, but it will take your mechanic about half

an hour. Checking the charge (with a hydrometer) and the water level and adding distilled water, if need be, is easy.

Time builds if the battery needs a charge and you have to unbutton the battery cover, schlepp the battery into the shop if need be, fool with the charger, and reinstall the battery. Might as well do it yourself.

If the battery needs a charge and you don't have a trickle charger, lug it into your mechanic's shop to be put on the charger. If you have never serviced the battery before, do it first under your mechanic's supervision.

$ **Savings** Do it yourself and save $10–$20 per servicing, which is mechanic's labor charge.

138 Bulbs and position lights

✻ **Action** FAR Part 43 permits "replacing bulbs, reflectors, and lenses of position and landing lights." Replace burned out bulbs yourself and save the labor charge (Fig. 5-2). Be sure to clearly understand what's involved. Some bulb changes require some quite complex dismantling of fairings and light housings. Get a mechanic to show you exactly what you need to do if there is any doubt. Also, always verify with the mechanic that you got the correct bulb.

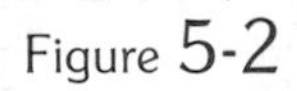

Figure 5-2

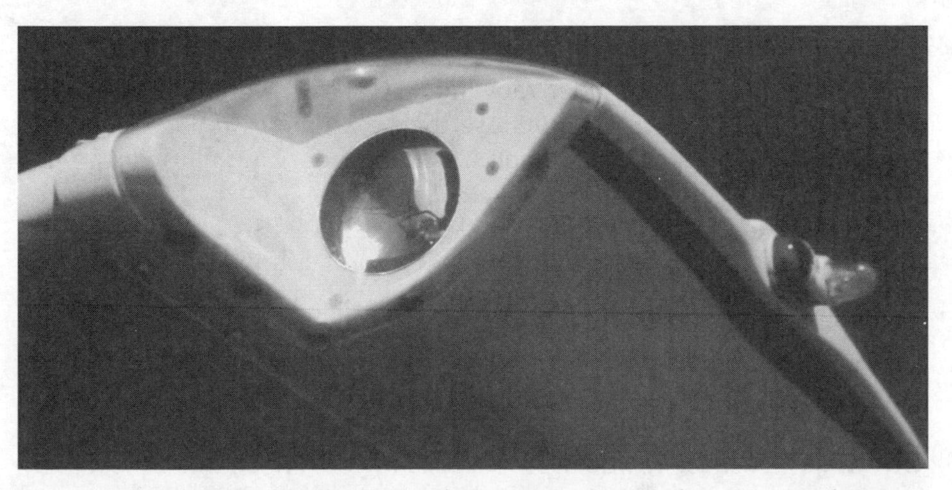

You can change these bulbs but make sure you know what you are doing.

$ **Savings** You will save \$10–\$20 per bulb change. You would likely be charged 15 to 30 minutes of labor for a bulb change.

139 Cowling replacement

✳ **Action** FAR Part 43 permits "replacing any cowling not requiring removal of the propeller or disconnection of flight controls." Is the cowling cracked and ratty beyond salvation? Replace it yourself. You can even paint the new one yourself, according to FAR 43.

Be careful, though, to be fully aware of the magnitude of the job you are taking on. Most cowlings are not the products of computer-aided manufacturing; hence, they rarely come ready to be snapped in place. A lot of fitting and hardware attachment might be required to get the job done. Seek your mechanic's guidance if you are inexperienced; always have the completed job checked out before you fly.

$ **Savings** You could save \$80 and up. It will most likely take a mechanic at least a couple hours or more to wrestle a replacement cowling in place. That can amount to a good chunk of change, depending on labor rates.

140 Elastic shock absorbers

✳ **Action** FAR Part 43 permits "replacing elastic shock absorber cords on landing gear." Many older ragwing airplanes have bungee-chord shock absorbers that eventually lose their elasticity. Replace the exhausted bungee chords yourself if you are bouncing all over the ground and the problem isn't you.

This is not the simplest operation and requires a bungee stretching tool. It is a time-consuming and specialized job, but chances are that if you fly a Piper Cub or any of the other golden oldies equipped with bungee-chord shock absorbers, you know how to do it, or you will be most willing to learn from a qualified mechanic.

$ **Savings** Because it would probably take your mechanic at least an hour to replace the elastic shock chords, you save \$40–\$80.

141 Fabric repair

Action FAR Part 43 permits "making simple fabric patches not requiring rib stitching or the removal of structural parts or control surfaces. In the case of balloons, the making of small fabric repairs to envelopes (as defined in, and in accordance with, the balloon manufacturer's instructions) not requiring load tape repair or replacement."

Stuck a rake through the wing? Patch the hole in the fabric yourself. Follow the fabric manufacturer's simple instructions. Some fabric suppliers might provide fabric repair kits.

Before undertaking fabric patching, carefully ascertain that no structural damage has occurred and that no rib stitching is required as part of the repair.

The typical procedure is to trim the tear neatly, usually into a circular or elliptical shape with no jagged edges. Cut a patch to overlap the edge of the damage by at least 2 inches all around. Apply butyrate dope to the surface being repaired to loosen and remove the previous coatings. When the surface is clean, coat it with dope and stick the patch in place. Smooth out the patch so that it is wrinkle free. Let it dry and apply several additional coats of clear dope, followed by silver dope, and finish off with colored dope matching the airplane's color. (If the airplane is painted with enamel, you will first have to gently sand off the paint down to the primer.) When the repair is completed, you can also touch up the paint according to FAR 43.

Warning: Fabric gets exposed to high-speed airflow (even 90 mph counts as high); therefore, it is essential that you use only FAA-approved materials and techniques when making even minor fabric repairs. Absolutely do not slap duct tape or any other unapproved material on a hole or a tear. The smallest tear could have potentially unsafe consequences if not properly repaired, and when you hear that depressing ripping noise as you merrily fly along, it might be a little late to become a believer.

$ **Savings** You would save $40–$80 because it will take your mechanic at least an hour or two to prepare and apply a fabric patch.

Fuel and oil filters/strainers—oil changes

✻ **Action** FAR Part 43 permits "cleaning fuel and oil strainers or filter elements." While FAR 43 does not explicitly authorize owner-performed oil changes, this is the main clause interpreted to allow such preventative maintenance. To clean the oil strainer and filter element, you have to first drain the oil.

Most filters are throw-away items, replaced, not cleaned. Replacement, too, requires draining the oil. Once you have drained the oil, cleaned the strainer and filter, or replaced the filter, you have to replace the oil, so, presto, you have completed an oil change.

A very important part of the oil change is safetying everything properly when the task is complete. Another section of Part 43 allows for the replacement of defective safety wiring and cotter keys. Because you made any safety wiring and cotter keys "defective" when you removed the strainers and filters, you are covered to replace them. Take great care to learn proper safety wiring technique.

A word of warning: An oil change, while not particularly complicated, is no small matter. The slightest error can have disastrous consequences. Each airplane has its own, specific oil change procedures. Learn them well under the guidance of a qualified mechanic before you perform solo oil changes.

You can also clean the fuel strainers. Wash them in clean fuel and blow-dry them. Take great care to safety them properly, following reinstallation. There is usually a main fuel screen and another fuel screen in the carburetor, if the engine is carbureted. Stay away from the fuel screen in the carburetor until a qualified mechanic has trained you thoroughly; carburetors are sensitive devices.

$ **Savings** Doing the oil change and related servicing could save you $40 and more because a typical labor charge for on oil change on a light single-engine airplane runs about that much.

143 Fuel line replacement

❋ **Action** FAR Part 43 permits "replacing prefabricated fuel lines." This section is primarily intended to allow the owner to change the flexible fuel lines in the engine compartment if they start becoming frayed. These lines (hoses, really) are made up to exact length, complete with self-contained connectors at both ends.

A word of warning: You better really know what you are doing before you take advantage of this one! You must be absolutely certain of having the correct replacement part. Verify it with your mechanic. A look-alike, but wrong part, or faulty installation could be catastrophic.

$ **Savings** You are likely to be charged at least half an hour of labor, if not more, for the replacement of these fuel lines. You save $20 and more.

144 Hose replacement

❋ **Action** FAR Part 43 permits "replacing any hose connection except hydraulic connections." This section allows the owner to replace the hoses running to the instruments, as well as the air hoses supplying air intake for heating and ventilating systems.

Be sure to replace hoses only with the appropriate replacement parts. The hoses serving the instruments, such as the vacuum pump, the oil pressure, fuel pressure, and manifold pressure hoses might be deliberately designed to a certain length in a particular airplane type and routed a certain way to avoid chafing and consequent failure. Reinstall the replacement hose exactly as the removed hose was installed. Pay particular attention to how the hose passes through the firewall and how the firewall opening is sealed.

Hoses connect in a variety of ways, all of them quite self-evident on inspection. Some hoses are clamped on, some are attached with a captive nut.

$ **Savings** You save $40 and up rather than paying for someone else's labor. Depending on the complexity of the routing, replacing a hose can take an hour or more of labor.

145 Hydraulic fluid

* **Action** FAR Part 43 permits "replenishing hydraulic fluid in the hydraulic reservoir." This is an easy task, but you must be careful to use the correct hydraulic fluid for your aircraft. The fluid used in cars is not the same as the one appropriate for aircraft, and different types of aircraft might use different fluid. Check your aircraft manual and verify the specs with your mechanic.

$ **Savings** You will save $5–$10 per replenishment. If the cowling has to be removed to gain access to the hydraulic reservoir, a bit more labor might be involved than you think; therefore, you would save that much more money.

146 Landing light maintenance

* **Action** FAR Part 43 permits "troubleshooting and repairing broken circuits in landing light wiring circuits." You can do several simple checks if a landing light does not work.

Check the bulb; you should be able to see if the filament is broken. Check any quick-disconnector fittings in the wires. Quick-disconnectors are found at places where, for maintenance access—such as the removal of the cowling—the wires need to be disconnected. Sometimes the disconnectors are reconnected improperly or inadvertently left unconnected.

Check the appropriate fuses or circuit breakers. If you know something about electricity, you can use a "hot light" or multimeter to run further simple circuit checks. If not, ask your mechanic to show you how.

A word of warning: Finding the problem does not necessarily mean that you can repair it. Given the complexity of the required repair, a mechanic might have to do it. Always check with your mechanic before you proceed to repair landing lights.

$ **Savings** Tracking down an electrical problem in the landing light circuitry can take as much as an hour. You save $20 and more.

147 Lubrication

* **Action** FAR Part 43 permits "lubrication not requiring disassembly other than removal of nonstructural items such as cover plates, cowlings, and fairings." Be sure to use the correct lubricant and lubricate only as frequently as required by the manufacturer. Check your aircraft manual.

$ **Savings** It can take a mechanic an hour or more to perform the required lubrication, including dismantling panels and cowlings to get to the parts to be lubricated. You could save $40 or more.

148 Magnetic chip detectors

* **Action** FAR Part 43 permits "removing, checking, and replacing magnetic chip detectors." Chip detectors find metal in the engine oil. This affects the operators of turbine-powered aircraft and all helicopters, including the most popular training helicopter, the Lycoming-powered Robinson R-22. Because turbine aircraft owners most likely can afford not to do their own maintenance, this tip is most relevant to owners of piston-powered helicopters.

$ **Savings** If this FAR applies to you, you can save about a half an hour to an hour's worth of labor: $20–$40 per servicing.

149 Maintenance—barter

* **Action** If you have any services or goods to offer that would be beneficial for your mechanic, make a barter arrangement for

maintenance. The most likely opportunities are for bookkeeping, accounting, and legal services, which a mechanic or FBO typically needs. Another excellent option is to barter flying time in your airplane for maintenance. Perhaps you can swap a specific item, such as a car you want to sell, for maintenance.

Whatever the exchange, agree to a mutually fair valuation, and spell out expectations on both sides in great detail to avoid potential misunderstanding.

$ **Savings** You might save hundreds of dollars. You and the mechanic save cash. You can also give each other a fairly generous, mutually acceptable discount because you both want something that would cost more if you had to buy it on the open market.

150 Maintenance—preapprove costs

✻ **Action** Preapprove all maintenance expenses. Get written estimates, and make it known in no uncertain terms that any additional expense must be specifically approved by you, ideally with your signature to avoid any misunderstanding that you or the mechanic might have when the bill is due. Never tell a shop or mechanic to "do whatever it takes to get something fixed."

You will save because there will be no unexpected surprises that might not have been entirely necessary to begin with. Also, the shop or mechanic will know that you are determined to control costs and will be less likely to subject you to a hard sell regarding discretionary maintenance.

$ **Savings** When you prevent unnecessary or unwanted maintenance, the savings can be several hundred dollars.

Mechanics—freelance

✻ **Action** Find a mechanic who is freelancing over and above a full-time job, and is willing to offer highly competitive labor rates. Such opportunities are quite common. Ask area airplane owners for

references. Some mechanics have a series of part-time jobs and might be in the market for free-lance work to fill the gaps. Some flying clubs can't pay high wages to a mechanic and openly permit free-lance work over and above the mechanic's club commitments.

Mechanics working for an employer can offer competitive rates on after-hours free-lance work because even though they price their services to you below the shop rate that you would normally pay, they get to keep it all, matching, or even exceeding their hourly wages in the shop.

A potential drawback to after hours free-lance work is that the mechanic might not be able to complete work on your airplane as promptly as you would like because you are second in line behind formal job commitments.

$ **Savings** You might realize a savings of $10 or more per hour. If the shop rate is $40 per hour and the free-lance rate is $30 per hour, your hourly savings of $10 represent a 25 percent discount.

152 Mechanics—shop

✳ **Action** Negotiate a discount with the shop in exchange for exclusive maintenance. Let the shop know that you are looking for the most economical source of parts.

Regarding a labor discount, how successful you are might depend on how anxious a shop is for business. Some shops, usually monopolies at an airport, might have more work than they can handle and have no reason to negotiate. Shops in intense competition with each other on one airport might be quite willing to negotiate.

Carefully evaluate over time how you are being treated. The vast majority of shops are highly respectable and honest; however, occasionally one comes along that routinely stretches the amount of labor billed, to make up for whatever discount you might think you are getting.

Regarding a parts discount, as long as you let it be known that you are on a tight budget, the shop should not have any problems with

finding you the most economical options for parts. You will have to pay the shop's markup, but you should specify the less expensive retreaded tires and overhauled accessories.

It helps to let the shop know in no uncertain terms that you are well-informed and expect the best deals.

$ **Savings** Most if not all of a 10–15 percent discount will be on labor. Part savings are realized with overhauled items instead of new ones.

153 Parts and accessories—overhaul kits

* **Action** If you can find a mechanic that is willing to let you buy parts and you pay only for the labor, purchase the necessary kit or kits to overhaul parts and accessories, then have your mechanic do the refurbishing. In some instances the mechanic's labor charge might negate most of the savings, but the option is well worth checking out. (Negotiate the circumstances, as well as the price, ahead of time. Never walk unannounced into an FBO or shop with a box of overhaul kits and say "Here, I already bought the parts, I'll pay you for the labor.")

$ **Savings** Perhaps \$40 or more could be saved. For instance, a vacuum pump costs \$159. The repair kit costs \$70, a difference of \$89. If your mechanic can overhaul the pump in an hour for \$40, you save \$49.

154 Preservative application

* **Action** FAR Part 43 permits "applying preservative or protective material to components where no disassembly of any primary structure or operating system is involved and where such coating is not prohibited or is not contrary to good practices."

This section's intent is to allow owners to apply grease and other preservative coatings on the fuselage to protect it from the elements. In practice it applies mostly to floatplanes, amphibians, and flying boats that require copious doses of grease and other preservatives on the floats, hulls, and associated bolts and wires for protection against the water's corrosive effects.

$ **Savings** You save $40 per application. It might take a mechanic an hour or so to apply the protective coating, depending on aircraft size.

155 Safety belt replacement

✱ **Action** FAR Part 43 permits "replacing safety belts." It is an easy job to replace the seat belts. Replacement is required when fraying appears anywhere on the belts. Most belts attach with simple bolts and self-locking nuts.

A word of warning: Make absolutely sure that the replacement belts are FAA-approved for your specific aircraft.

$ **Savings** In most light aircraft, replacement of the belts takes about an hour to an hour and a half: $40–$60 saved.

156 Safetying

✱ **Action** FAR Part 43 permits "replacing defective safety wiring or cotter keys." Safety wiring and cotter keys are required on many parts of the airplane. Examples are fuel and oil filters, turnbuckles, and some bolted components.

Safety wiring has to be done in a particular way, which has to be learned.

The idea is that when the safety wire is in place, the wire should be routed to be pulled tighter by any tendency for the safetied object to become loose by any unscrewing or rotation due to vibration. Ask your mechanic to teach you proper safety wiring techniques.

The ends of cotter pins must be tightly bent flush with the nut following installation.

$ **Savings** Expect to save $5 and up. Safetying of individual items doesn't take long, but nevertheless contributes to total savings because safetying is an integral part of another job, such as changing the oil and cleaning the screens.

157 Seat replacement

⁕ **Action** FAR Part 43 permits "replacing seats or seat parts with replacement parts approved for the aircraft, not involving disassembly of any primary structure or operating system."

This section allows the owner to take the seats in and out during maintenance, such as the annual, and to replace any broken or malfunctioning seat components. Take great care to use only FAA-approved seat parts appropriate to your specific aircraft.

$ **Savings** It takes a mechanic approximately 15–30 minutes to remove and reinstall the seats in the average four-seater: save $10–$20. Seat repair might take longer.

158 Seat slipcovers

⁕ **Action** If the seat fabric deteriorates but you want a less expensive option to reupholstering the seats, make seat slipcovers instead. Slipcovers are easily sewn and are inexpensive. Usually the two front seats wear out first. Matching slipcovers restore the ambiance of the interior and can postpone the need for a complete reupholstering for years.

$ **Savings** You save $150–$250 per seat in comparison to upholstering.

159 Shock struts

⁕ **Action** FAR Part 43 permits "servicing landing gear shock struts by adding oil, air, or both." Fully servicing air-oil (oleo) shock struts is quite an involved and potentially messy process for the inexperienced person. The airplane has to be jacked up to service the main gears. The nose gear has to be lifted up to be serviced by weighing down the tail (Fig. 5-3).

Figure 5-3

Follow the manufacturer's instructions and FAR Part 43 to service the struts.

The air has to be let out of the strut, the valve removed, oil pumped into the strut, and the valve reinstalled. (Be sure to get the appropriate oil for the struts. Check the aircraft manual for the specs.) The airplane then has to be put back on its gear and the strut has to be inflated with compressed air to the appropriate level.

Servicing only with air is easier, with the airplane on its wheels. Compressed air is pumped into the air valve to inflate the strut to the required level. The wing should be gently shaken while the strut is being inflated to assure equal inflation.

In any case, get a mechanic to teach you how to service the struts before you do it solo.

(Regarding the strut seal condition, if the struts are getting low, it is most likely because of leaking strut seals, so have them checked. Only a qualified mechanic can change the seals. To minimize strut seal deterioration, wipe down the struts after every flight.)

$ **Savings** You save $80–$120. It can take two hours or more of labor to fully service all the struts. Adding air is less time consuming.

160 Shock struts—cleaning

* **Action** Wipe the shock struts clean after every flight. Any dirt and grit on the shock struts grinds away at the strut seals and eventually the seals deteriorate and must be replaced (Fig. 5-4).

By simply keeping the struts clean, you might be able to not only delay, but avoid altogether the need for strut seal replacement while you own the airplane.

$ **Savings** If you can avoid replacing the seals, you can expect to save as much as $100 per strut.

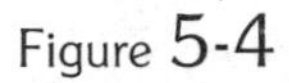

Figure 5-4

This dirty strut is heading for a strut seal change. A simple wipe down after every flight significantly prolongs the life of strut seals.

161 Spark plug maintenance

* **Action** FAR Part 43 permits "replacing or cleaning spark plugs and setting of spark plug gap clearance." This authority is useful when you are faced with fouled spark plugs, or at annual time. You must have the correct spark plug wrench and gapping tool, and must know how to use them.

Spark plugs are easily damaged. It is, therefore, especially important to pull the ignition harness leads straight out, hold the wrench correctly to put force on the plugs at exactly right angles, and pull out the plug precisely along its installed alignment.

Keep track of which cylinder each plug came out of because the condition of the plug's firing end tells you a lot about engine performance. Briefly: Normally worn plugs should be free of oil and

soot; black soot is a sign of an overly rich mixture; an exceptionally clean firing end might mean that the cylinder is running too hot; oil indicates a worn ring or valve guide.

To clean the plugs you have to remove any lead and carbon deposit from the firing cavity (scrape out with a sharp pick), lightly sandblast the plug, and clean the terminal sleeve (sometimes referred to as the cigarette because of its white ceramic appearance) with an appropriate solvent. You then gap the plug and are ready to reinstall it. If you have a tester, you should first check to see if the plug fires.

Reinstall the plugs as carefully as you took them out to prevent damage.

All well-equipped shops have spark plug cleaning machines that use high-frequency vibration to clean the plugs and do a much better job than the human hand. Arrange access to a machine for best results. (Refer to the next tip, which examines reconditioned spark plugs.)

A word of warning: While spark plug servicing and replacement is quite straightforward, the plugs can be easily damaged or improperly gapped. Make it a point to thoroughly learn from your mechanic how to service spark plugs before you do it solo.

$ **Savings** You could save $60 or more. It takes most mechanics at least an hour and a half, or longer, to take out, service, and reinstall the eight spark plugs of a typical four-cylinder aircraft engine.

162 Spark plugs—reconditioned

* **Action** Have your spark plugs reconditioned, or purchase reconditioned plugs instead of buying new plugs. Reconditioning consists of cleaning the plugs with a machine, gapping them, electronically testing them under pressure, diagnostically evaluating them, repainting them, and equipping them with a new gasket. Any well-equipped shop can recondition spark plugs.

Reconditioned plugs are also available for sale for considerably less than the cost of new plugs, although you might be taking a chance. It is better to have your plugs reconditioned.

$ **Savings** Expect to save $10 per spark plug, $80 for a set of eight. Typically, a reconditioned spark plug costs approximately $3 compared to a new plug for $13.

163 Tires

✻ **Action** FAR Part 43 permits "removal, installation, and repair of landing gear tires." This section allows the owner to remove the wheel, change tires, and repair flats. Servicing tires is quite simple, but actual repair will most likely be beyond the capabilities of most owners (Fig. 5-5).

Figure 5-5

Improper inflation reduces tire life by as much as 50 percent.

To remove a tire from the wheel rim you best have a special tool called a bead breaker. You will also need a jack (and know how to jack up the airplane properly). On most light aircraft you will also have to unbolt the brake assembly before you can take off the tire. Have your mechanic show you how to change and service tires before you do it on your own.

Be careful with high-pressure tire assemblies that can be explosive if mishandled. High-pressure tires are rare on light aircraft, but are encountered occasionally.

$ **Savings** You could save \$20–\$30. It rarely takes a mechanic more than a half an hour to replace a tire.

164 Upholstery work

✳ **Action** FAR Part 43 permits "repairing upholstery and decorative furnishings of the cabin, cockpit, or balloon basket interior when the repairing does not require disassembly of any primary structure or operating system or interfere with an operating system or affect primary structure of the aircraft."

This section allows the owner to be an interior decorator, as long as the cited restrictions are observed and the material is approved for aircraft. For owners on a budget, the best course of action is to install a prefabricated interior kit, such as the popular Airtex line.

$ **Savings** The savings can be significant because upholstery work, especially custom work, is very labor intensive: more than \$1,000. If you install a \$1,500 interior yourself instead of paying \$3,000 for a custom job, you save \$1,500. (Refer to Tip 81 in chapter 3 regarding upholstering kits vs. custom work.)

165 Wheel and ski change

✳ **Action** FAR Part 43 permits "replacing wheels and skis where no weight and balance computation is involved." Owners of ski planes who want to use skis and wheels interchangeably benefit from this authority.

$ **Savings** Savings are considerable in comparison to having a mechanic do it: $80 or more per change. A mechanic can easily take about two hours to switch skis for wheels or vice versa.

Wheel bearings

✻ **Action** FAR Part 43 permits "servicing landing gear wheel bearings, such as cleaning and greasing." When a wheel is off, such as for a tire change, the bearings should always be greased. Bearings should also be greased periodically, according to the aircraft's lubrication schedule. Ask your mechanic to show you how to perform this simple operation.

It is a bit of a production to take the wheel off; the airplane has to be jacked up and most likely the brake assembly has to be unbolted.

$ **Savings** Expect a savings of $10–$20 per wheel. It only takes a few minutes to grease the bearings, but more effort to remove and reinstall the wheel.

Window replacement

✻ **Action** FAR Part 43 permits "replacing side windows where that work does not interfere with the structure or any operating system such as controls, electrical equipment (and the like)."

Windows eventually get crazed, or they might get damaged. Replacement is labor intensive, providing a good opportunity for saving if you can do the job yourself. (Be sure to do only side windows, as authorized). It can, however, be a complicated job for the novice. Don't bite off more than you can chew. A good compromise is to do it under the supervision of your mechanic. You may do the monotonous work; the mechanic can do the fitting and provide guidance.

$ **Savings** You could save $100 or more. Changing a window can take about 2.5 hours.

6

Major overhaul and repair

 Accessories—new vs. overhauled

⁕ **Action** Send out malfunctioning accessories for overhaul, or exchange them for overhauled accessories instead of purchasing brand-new accessories. Professionally overhauled accessories function just as reliably as new ones, and usually also carry some form of warranty. Often the key parts of many overhauled accessories are new components.

The FAA authorizes repair stations to perform the overhaul of various instruments and accessories. Make sure that you are dealing with an authorized facility. The overhaul information on some instruments and accessories is evidenced by a yellow tag, a practice that has led to them being referred to as "yellow-tagged" items. A good source of information on the availability and cost of overhauled items is *Trade-A-Plane*.

The less expensive overhaul option is to have the accessories and instruments sent out to be overhauled. The drawback is having to wait as much as several weeks for them to be returned. The more expensive but quicker option is to exchange your malfunctioning accessories and instruments for overhauled replacement units.

$ **Savings** You should save as much as 50 percent compared to the cost of new items.

169 Accessories—when to overhaul?

✲ **Action** Overhaul accessories only when they begin to malfunction. If you are on a budget, there is no sense in overhauling an auxiliary fuel pump that works perfectly well just because the engine is being overhauled and it makes sense to get everything done all at once. Vacuum pumps, for example, easily last about 500 hours. It would, therefore, be extravagant to overhaul one that has 300 hours on it because the engine happens to be undergoing overhaul.

If you fly IFR and are concerned about safety, be sensible and invest in a standby vacuum pump (regardless of being on a budget), but don't overhaul the main vacuum pump before its time.

The effect of overhauling items prior to their time is that they add to the per hour cost of flying, because their cost is spread over fewer flying hours.

$ **Savings** Can be several hundred dollars or more cumulatively. Take an auxiliary fuel pump, a vacuum pump, a starter motor, and two magnetos. Suppose their cumulative cost including installation was $3,000. Assume that at the time of engine overhaul each of these accessories has 400 hours and is in perfectly good working order to run another 300 hours longer.

If the accessories are overhauled at this time, their cost per flying hour would have been $7.50 (3,000 ÷ 400). If they are left alone for another 300 hours, their cost per flying hours would have been $4.28 (3,000 ÷ 700), a per hour savings of $3.22, or an additional $966 worth of use during the extra 300 hours (3.22 × 300).

In real life, of course, some accessories will last longer than 700 hours, some fewer, but if the costs are averaged, similar savings could result.

170 Engine overhaul—"backyard" option

✻ **Action** Opt for a less expensive "backyard" overhaul, rather than an overhaul by a big engine shop. A backyard overhaul is a major overhaul done by your local friendly mechanic. At face value this suggestion might sound like appalling heresy. Doesn't everyone have a favorite backyard overhaul horror story? Perhaps, but objective reality is, as usual, not black and white, but many shades of gray.

Done by the right mechanic, the backyard job is as good or better than the assembly line engine shop variety and can cost less, to boot. There is a vast army of individualistic master craftsmen out there ready to do the job.

Understanding a backyard overhaul can go a long way to convince you to at least seriously consider one. If properly done, the backyard overhaul gives the engine the same pampered treatment as any other overhaul. Cylinders and crankshaft are replaced by factory-new parts, the crankcase is sent out to an appropriately equipped facility to be Magnafluxed for any cracks, the desired tolerances are strictly observed, a new wiring harness and spark plugs are installed, all accessories are overhauled as needed, and so on. The difference is that the backyard craftsman may do all the labor at a lower rate than what you might be charged otherwise.

A word of warning: The crucial element to the success of the backyard overhaul is finding the right craftsman for the right price. This option is not for the faint of heart or the uninformed. You have to have a thorough understanding of overhaul options, and a lot of experience selecting the right person for the job. You had best thoroughly know the mechanic, plus the mechanic's work and abilities from firsthand experience. When the mechanic has your full faith, the two of you better sit down and establish exactly what the overhaul will include, to what tolerances it will be performed, and how much it will cost.

There are two potential drawbacks to the backyard option. First, it might take quite a bit longer than an assembly line shop overhaul. Second, when you offer the airplane for sale, it might make lookers wary.

$ **Savings** You might save several thousand dollars, depending on the engine.

171 Engine overhaul—chromed cylinders

✻ **Action** If you do not fly your airplane frequently, buy an airplane with chromed cylinders or get them chromed at the next major overhaul. Opinion on the value of chromed cylinders is mixed. It is a fact that because the chrome hardens the metal, the chrome prevents the pistons from seating as snugly as they would otherwise. As a result, engines with chromed cylinders tend to consume somewhat more oil.

If an airplane is flown about 15 hours per month or more, chromed cylinders are probably unnecessary. The pistons will seat more snugly as the two softer metal surfaces work against each other during the frequent flights and moisture will be burned out of the engine with sufficient regularity to prevent a rust problem from developing.

However, if the aircraft flies fewer than 15 hours per month, the pistons won't get a sufficiently steady workout to seat as tightly as they could, leaving an opening for accumulated moisture to cause a problem. In this case, chromed cylinders will greatly reduce the chance of a rust problem and will in most cases prolong engine life.

$ **Savings** Hundreds or even thousands of dollars could be saved, depending on the prolongation of engine life and on the overhaul costs. Potential savings result from the greater number of flight hours obtained from an engine due to chroming.

Consider the following scenario: The cost of an overhaul is $15,000; TBO is 2,000 hours; a chromed engine made it to TBO while the nonchromed engine had to be overhauled at 1,500 hours, the per-hour cost of the chromed engine is $7.50, compared to $10 per

hour for the nonchromed engine. This is a saving of $2.50 per hour. Another way of looking at it is getting an extra $1,250 worth of flying out of the chromed engine (2.50 × 500).

172 Engine overhaul—factory vs. nonfactory

* **Action** Get the engine overhauled by a reputable independent overhaul facility rather than by the engine factory. The most expensive engine overhaul option is to have it done by the engine factory. Less costly alternatives are licensed FAA repair stations, maintenance shops, and independent mechanics.

Licensed repair stations have been inspected by the FAA and authorized to sign off on major work performed by the repair station. Maintenance shops and independent mechanics, on the other hand, have to have such work signed off by an IA. (Many of them have an IA on staff.)

There are differences in the tolerances to which different facilities may officially overhaul engines. Only the factory can "zero time" an overhauled engine by overhauling it to factory-new tolerances, throwing away the old engine logs, and issuing new logs with zero time.

Even though they often perform overhauls to factory-new tolerances, repair stations and independent mechanics can document an overhaul only to *service limits*. Service limits mean that the engine components are within tolerance to be in service, but might not be up to factory-new standards. The old logs have to be retained, and the engine cannot be "zero timed."

In practice, nonfactory overhaul facilities will perform the work to whatever standards you specify and document it in an invoice to you, regardless of the limitation imposed on engine log entries. The most important thing from your perspective is to fully understand what is included in the overhaul you are accepting. Don't be shy to demand explicit detail and comprehensive explanations.

In most cases, the most practical and cost-effective solution is to have your local shop send out the engine to a nonfactory engine overhaul specialty shop, and reinstall it for a preapproved fee, following the overhaul.

$ **Savings** You can save up to several thousand dollars. A nonfactory overhaul (performed to standards that are equal to a factory overhaul) will usually cost 15–20 percent less than the direct factory overhaul cost.

173 Engine overhaul—to what tolerance?

✳ **Action** Under certain circumstances consider overhauling the engine to service limits instead of factory-new tolerance. Two basic overhaul alternatives are factory-new tolerance and service limits.

Under a factory new tolerance, all engine components that do not meet tolerance limits of factory new components are replaced. In essence, the overhauled engine is as good as a brand new engine straight from the factory.

A description of overhauling to service limits requires an understanding of service limit tolerances. When an engine enters service and experiences wear and tear, its original tolerances deteriorate. When tolerances deteriorate to certain limits, the engine no longer meets the tolerances required to remain airworthy. These limits are called service limits. The main requirement of the overhaul to service limits is that all engine tolerances be above this limit. From a practical standpoint, this means that only those parts that are below service limits need to be replaced at overhaul time. If, for example, the crankshaft is examined and found to be within service limits but not factory-new condition, it need not be replaced. An overhaul to service limits can be much less expensive than an overhaul to factory-new tolerances.

A word of warning: An overhaul to service limits can be very minimal in the sense that if a component barely meets service limits it may be left alone. If shortly after the overhaul the component's tolerance falls below limits, it must be dealt with at potentially great

additional expense. One could end up being penny-wise and pound-foolish. An overhaul need not always be done to factory-new tolerances, but great common sense has to be applied to decide what tolerances should be acceptable on the spectrum between factory-new tolerances and service limits.

It is generally shortsighted to overhaul a runout engine to barely above service limits. This option makes the most sense when a generally healthy and reasonably low-time engine has to be overhauled because of some specific reason such as a prop strike or crankcase crack.

$ **Savings** Potentially several thousand dollars could be saved by careful selection of the overhaul technique. The savings come mostly in parts costs. There is little scope for labor savings because the same amount of work has to be done to dismantle an engine for overhaul, then examine the components, and finally reassemble the engine, regardless of how many components are actually replaced.

174 Engine overhaul—when to overhaul?

✳ **Action** Overhaul the engine when it really shows the kind of wear and tear that warrants an overhaul, not when it reaches recommended time between overhaul (TBO).

TBO is a recommended limit, not a legal limit. If the compression is good and the engine is in generally good condition, there is no reason not to go beyond recommended TBO. It is not unusual for an engine to last several hundred hours beyond recommended TBO. If it goes only an extra 200 hours (10 percent of a recommended TBO of 2,000 hours) it might mean an extra season or two of flying for you.

Do not exceed recommended TBO without an especially thorough mechanical inspection by a qualified mechanic. Refer to chapter 4 and tip 125 regarding exceeding TBO.

$ **Savings** Expect to save hundreds of dollars by getting more value for your money by getting the most use out of an engine prior to overhaul.

175 Engine purchase—new vs. overhauled

* **Action** When you have to purchase an engine, instead of overhauling the one you have, buy an overhauled engine instead of a factory-new engine. Brand-new engines are notoriously expensive and the prices are rising all the time. Properly overhauled engines work just as well for just as many hours for far less expense.

$ **Savings** Potential savings are in the thousands of dollars. The price of overhauled engines is commonly as much as 35 percent or more below the price of factory-new engines.

176 Fabric—replacement

* **Action** When it comes time to replace the fabric on your airplane, seek out a fabric specialist at some rural airport even if it is some distance from your home base. Don't pull into your big, local all-purpose maintenance facility.

Fabric replacement is a specialty and is not in great demand these days. The few craftsmen who specialize in fabric are more likely to be in rural locations where overheads are low, resulting in savings (usually passed on to the consumer to a certain extent) compared to the multipurpose shops in busy population centers. Additional savings come from the ability of the specialty shop to complete the job with less labor (because of a higher level of experience) than the big, all-purpose shop.

$ **Savings** You can expect to save several hundred dollars compared to the big, all-purpose shop.

177 Major repairs and alterations

* **Action** In the unlucky instance of having to have major repair performed on your aircraft or if the aircraft needs major alterations, have the work performed by an independent mechanic specializing in the particular repair required, rather than a big shop.

Many mechanics willing to moonlight also specialize in some aspect of repair work, such as sheet metal work. Their hourly rates are usually very reasonable compared to shop rates, and they are willing to take on major repairs or alterations as a project.

While the savings are usually considerable under such an arrangement, be aware that the job might take quite a long time if the mechanic also has other commitments; however, the extra time might be well worth the savings.

$ **Savings** Several hundreds of dollars can be saved, depending on the job.

178 Owner work under A&P supervision and sign-off

✻ **Action** If you are mechanically inclined, but do not have an A&P license, find a mechanic or shop that allows you to work on the aircraft under supervision. As long as a properly certified mechanic supervises your work, inspects the results at the appropriate stages and signs off on the job, you can do many tasks yourself. The savings on labor costs can be significant.

Your role can be as simple as taking apart components for overhaul or repair, or a lot more complex, depending on your abilities. An added benefit of owner participation is learning a lot about your airplane.

Safety point: When participating in overhaul or repair work, make absolutely sure that your mechanic conducts the appropriate inspections of your handiwork. Do not overreach and be especially careful not to violate FARs.

$ **Savings** Expect to save hundreds of dollars. You would save $400 if you did 10 hours of work and apply a mechanic's $40 per hour shop rate.

179 Repainting

✳ **Action** Shop around carefully for an aircraft paint shop when it comes time to repaint your airplane. There are several potential saving opportunities. First, shops in rural locations might be extra competitive because of lower overheads.

Second, depending on local and state regulations, certain shops might operate under less costly environmental requirements than others and can pass some of these savings on to you.

Third, certain shops might allow you to participate in selected aspects of the job such as stripping and masking surfaces or removing control surfaces, fairings, gear doors, and the like. You save on labor costs.

Even though you are hunting for a bargain, be cautious about signing up for a paint job that is below the standards you are looking for. Establish the quality you want, see examples of the various facilities' work, and then find the best bargain within the parameters you set (Fig. 6-1).

Figure 6-1

Civilian and military paint schemes are equally appropriate for the Navion. Find a paint shop that lets you help out with the repainting to cut costs.

$ **Savings** Shopping for location and perhaps doing some of the labor could save you several hundred dollars.

180 Top overhaul vs. major overhaul

✻ **Action** If your airplane's engine experiences compression problems before recommended TBO, consider a top overhaul instead of a premature major overhaul. The top overhaul is much less expensive; it might yield enough additional hours to TBO to be financially worthwhile.

Eventually you have to bite the bullet and spend the money for the major overhaul. But depending on when you need an overhaul and how much you fly, the top overhaul can be a much less costly alternative that keeps you flying for years.

$ **Savings** The top overhaul alternative can save several thousand dollars. Suppose you fly 100 hours a year and the compression goes on several cylinders at 1,500 hours. A $3,000 top overhaul that yields 500 additional hours to TBO means flying 5 more years for a lot less money than a major overhaul, which might cost at least $15,000.

7

Learning to fly

Flight instruction—bulk time purchase

✻ **Action** Purchase dual time in bulk to benefit from a discount. Just as they sell bulk time to licensed pilots, most FBOs and flying clubs also sell bulk time to student pilots. A typical block is 25 hours. The more time you purchase at once, the greater the discount.

If bulk time is apparently unavailable, suggest it to the school. You might be offered a personal deal or bulk time might be formally instituted at the school.

A word of warning: It might be difficult to accurately assess the financial health of the school, so exercise some caution in how much money you pour into bulk time.

$ **Savings** A 10–20 percent saving in comparison to regular rental rates is typical. If the regular dual rate for a Cessna 150 is $45 per hour and you get 20 percent off for buying 75 hours of block time, you save $675.

Flight instruction—free-lance instructors

✲ **Action** Get a free-lance instructor rather than one through an FBO if you have an airplane or access to one independent of an FBO. FBOs charge a fair amount for instruction. They have to make their margin and meet overhead before giving the rest to the instructor. A free-lance instructor contracting with you directly can afford to charge less than you would pay the FBO, but more than the instructor would get from the FBO. You and the instructor both come out ahead.

$ **Savings** It is not unusual to save $10 per hour. For example, 40 hours of instruction from a free-lance instructor at a savings of $10 per hour will save you $400.

Flight instruction—package courses

✲ **Action** Take a total-immersion accelerated package course instead of working toward a rating part-time, only flying hour or two per week. The price of most accelerated courses is "guaranteed"—you pay a fixed price for the rating even if you need a few extra hours to complete it.

The guaranteed price is usually competitive in comparison to the cost of the training if flying part time. The school can afford to offer a good price because it knows that you are likely to qualify for the rating in minimum flight time given your total immersion in flying. On the other hand, if training part-time you are likely to require considerably more flight time for the same rating due to interruptions of your weekend flying.

A word of warning: Carefully check out the reputation of the operators offering accelerated courses and be wary of requests for substantial payments in advance.

$ **Savings** You could save $1,000–$2,000 depending on the rating.

 184 Ground school—accelerated programs

❊ **Action** Take a weekend accelerated ground school instead of studying individually with an instructor or in class that meets once a week. The accelerated ground schools include the books, examiner's fees, and taking the written at the end of the course. If you fail, most companies allow you to repeat the course until you pass. These programs can save money for those students who lack the discipline for self-study and whose other alternative is fairly expensive hourly individualized instruction.

$ **Savings** Depending on how much the course costs and how much individual instruction you require as an alternative, you can save as much as $100.

 185 Ground school—home study

❊ **Action** Study at home, independently, with minimal instructor assistance. This is the least expensive ground school option, if you can hack it. Get an inexpensive self-study ground school package and hit the books. It is really quite easy if you have the discipline.

The packages include all sorts of practice exams that you can take as often as you like. Consult your instructor only to clarify confusion. When you are ready, submit yourself to a good grilling by your instructor and a sign-off to take the written.

Home study software is becoming increasingly popular. It is more pricey than books, but can be less costly than some formal ground schools. It is the most flexible study aid of all the options. Its interactive features make comprehension a snap. Practice exams can be taken infinitely and are graded instantly.

$ **Savings** Home study might save as much as $200, compared to the alternatives.

186 Ground school—sharing home study materials

* **Action** Get together with a group of student pilots and share the costs and use of home study ground school materials such as software and videos. Observe copyright requirements, perhaps pass the material around from student to student if copying is unauthorized.

$ **Savings** Suppose that four people pitch in $50 for a $200 ground school software package. Each person saves $150 compared to the alternative of buying it alone.

187 Ground school—rent video courses

* **Action** Rent ground school video courses from an FBO or pilot shop instead of buying the course. This option restricts viewing time, but rental rates are sufficiently competitive to enable you to rent the courses for an ample time and still come out ahead. Plan your rental carefully to squeeze the maximum benefit out of the least amount of rental time.

$ **Savings** Expect to save as much as $100. Consider a six-tape $200 video course that rents for $2 per tape per day. Complete each tape in 8 days and pay $16 per tape; renting the six tapes will cost $96. You have saved $104.

188 Learning to fly—flying club

* **Action** Join a flying club and learn to fly there, rather than learning at an FBO's flight school. Rental costs will be lower and eventually the club's annual dues will be absorbed because you will fly many, many hours as a student pilot. (Weigh all the other pros and cons of flying clubs and FBO's before you decide on the club strictly for the savings.)

$ **Savings** A private pilot certificate might cost $1,500 less at the club. If it takes 75 hours to get the private certificate and you are saving $20 per hour at the flying club, you will save $1,500 in comparison to the FBO flight school alternative.

 Learning to fly—IFR rating

✻ **Action** Employ all cost-cutting strategies to minimize the cost of the IFR rating. In addition to general measures to cut the cost of training discussed elsewhere, there are several savings techniques specific to the IFR rating.

First, if you are renting a trainer, get as much training as possible in a less expensive, VFR aircraft. Most of the basic hood work, as well as any instrument work requiring one VOR, can be done in a VFR aircraft that has a single navcom.

Second, if you have your own aircraft, even if it is only VFR, train in that airplane as much as possible. If you are going to upgrade a VFR airplane to IFR capabilities, or if you are going to buy an IFR airplane, do so before you start training and train entirely in an IFR airplane.

Third, you are allowed to log as much as 20 hours of IFR training in an FAA-approved procedures simulator, such as a Frasca. Take full advantage of this less expensive training option. If you fly several times a week and are of average ability, you should be able to qualify for the rating in a little over the minimum flight time required, even though you do half of it in the simulator.

Finally, train for the IFR rating in one concentrated burst of effort. To a greater extent than for any other rating, consistency is key to getting the IFR rating in the minimum time required. IFR skills erode very quickly, especially at the outset, and frequent breaks of a few weeks between training flights will practically wipe out anything learned in the previous couple of sessions.

$ **Savings** Expect to pocket several hundred dollars when every possible saving technique is applied to an IFR rating.

 Learning to fly—airplane ownership

✻ **Action** Buy your own airplane to learn to fly. Radical as it might sound, you can save money by buying a budget trainer, hiring an

independent instructor, and selling the trainer after you obtain the certificate. If you select a trainer with a high resale value, after 75–100 hours of training you can most likely sell it for the price you paid.

If you were planning to buy an airplane after learning to fly, buy it at the outset and save money learning in it. If the tab for buying the airplane is too hefty, consider buying a trainer in partnership with other student pilots.

Evaluate this option thoroughly to get a realistic measure of potential savings. An excessive number of hours—beyond what is necessary for a certificate—might be necessary to make this option economical. If you plan to fly more hours anyway, this might be the best solution; otherwise you are better off at the FBO or flying club.

$ **Savings** Potentially hundreds of dollars can be saved, depending on a variety of factors, such as rental rates, ownership costs, flight hours per year, and resale value.

191 VFR vs. IFR

✳ **Action** Don't get an IFR rating if you realistically will not have much use for it. The IFR rating is expensive. Its main purpose is to facilitate transportation. If you are on a budget and don't have reason to go on frequent cross-country flights in poor weather, maintaining IFR currency might be more trouble than the rating is worth. Many pilots have every intention to use the IFR rating when they set out to get it, only to find it impractical to maintain proficiency.

A clear benefit of the IFR rating—use it or not—is that the training will make you a better pilot. But the expense of getting the rating and maintaining it is going to put a tremendous dent in your bank account for something you might not use very much.

$ **Savings** You could save $3,000–$4,500, the typical cost of an IFR rating.

Take advantage of the FARs

Warning: FARs change over time. It is every pilot's responsibility to keep informed of all regulations by consulting the current FARs directly. Many of these tips are based on postponing renewal of FAR-mandated items by a brief time period beyond the expiry of the item. When anything is postponed beyond expiration, do not fly as pilot in command, or operate any aircraft until everything is legal—including the pilot.

192 Annual inspection—extra month

* **Action** Postpone an annual inspection to the month immediately following expiration and gain an extra month toward the next renewal date. Under FAR §91.409 an annual inspection is valid until the last day of the 12th month following the the month in which the annual was performed. For example, an annual performed anytime in January 1994 is valid until January 31, 1995. Arrange for the

new annual to be completed on February 1 or later. Your next annual will then not be due until the last day of February in 1996, at which time you can repeat the "extra month" process.

$ **Savings** Typically you could save $50 or more per year. The extra month saves 1⁄12 of the cost of the annual inspection. If the inspection is $750, you save $62.50 for the year. Keep up the practice for 12 years and cumulatively earn a free annual.

193 Biennial flight review—extra month

✳ **Action** Postpone your biennial flight review to the month immediately following expiration. A biennial is valid until the last day of the 24th month following the month in which it was performed; therefore, if you renew on the first day of the month following the month in which it expires, you gain an "extra month." Repeat the process at every biennial.

$ **Savings** Typically expect to save $20 and more per biennial. Biennials are expensive by the time you are done with ground instruction and flying time. Suppose that it cost you $600 for a biennial all-in (the cost of aircraft rental, instruction, preparatory flights, etc.). The "extra month" technique saves you 1⁄24 of the biennial cost, or $25. This is not a great sum, but the timing to make the "extra month" work is easily arranged, and $25 will buy you enough avgas to fly an Arrow 2.5 hours.

194 Medical—class

✳ **Action** Get a medical appropriate only to the level of flying you do, regardless of pilot's certificate held. The higher the class of the medical, the more expensive it is and the more frequently it has to be renewed. If you have a commercial certificate but don't fly commercially, why have a more expensive Class II medical that has to be renewed in 12 months when a private pilot's Class III medical costs less and has to be renewed only every 24 months?

$ **Savings** Expect to save as much as \$500. Say, the Class II medical costs \$100 and has to be renewed every year; the Class III medical costs \$75 and has to be renewed only every 2 years. For the 2-year period, the savings are \$125 [(2 × 100) – 75]. Considerably more is saved if the Class III medical is the alternative to a Class I medical.

195 Medical—extra month

✻ **Action** Postpone your medical examination to the month immediately following expiration. Under FAR §61.23, medicals are valid until the last day of the month in which they expire. A Class III medical obtained in May 1994 is valid until May 31, 1996. Arrange for a medical examination on June 1 or soon thereafter. The new medical will expire on June 30, 1998, and you can repeat the "extra month" process. This tip is applicable to all medical classes. Do not fly until an expired medical has been renewed.

$ **Savings** You can save \$3–\$15 per medical period. Savings calculation: A Class III medical examination costs \$75. Because the medical is valid for 24 months, the "extra month" savings is 1/24 of the examination fee, which equals \$3. Several dollars isn't much, but it will pay for a couple of sodas at a nearby fly-in restaurant. Savings on the more expensive, higher class medicals will be more.

196 Static system and altimeter check

✻ **Action** There are two opportunities to save on static system and altimeter checks. First, disregard the check if it is not legally required for the type of flying you do. If you fly only day VFR and only in airspace that does not require an altitude encoding transponder, there is no need for the expensive checks. Performing the checks is an excellent precaution, but not a requirement for day VFR.

Second, postpone the required checks to the month immediately following expiration and gain an extra month toward the next renewal date. (When anything is postponed beyond expiration, do not operate any aircraft until every component is "legal"—including the

pilot.) Under FAR §91.411, static system and altimeter checks are valid until the last day of the 24th month after the month of the check; schedule the next check early in the 25th month and do not fly until the check is complete.

$ **Savings** Expect to save $10–$15 from this "extra month" technique. The figure might rise to $300 or more if you can forego the checks. Gaining the extra month saves 1/24 of the cost of the checks. If the checks cost $300, your savings from the extra month are $12.50. Not much, but every penny counts.

Transponder

✻ **Action** Do not purchase a transponder if you fly only VFR in airspace where a transponder is not required. This tip applies to a narrow segment of the pilot population because transponders are practically universal in the United States nowadays and are an important safety device, even for VFR aircraft; however, if you fly only local VFR on a budget in airspace not requiring transponders, you might want to forego the expense.

$ **Savings** You will save about $1,000.

9

Don't neglect these tips

198 Actively oppose antiaviation groups

✳ **Action** Actively oppose any antiaviation group intent on restricting our ability to fly. The freedom to fly is being constantly challenged in a variety of ways at various levels. Developers build houses on the vacant lots next to an airport that has been there for years, then push sales of the houses to unsuspecting buyers, As soon as the developers have cashed in, the new homeowners start a movement to impose noise restrictions. Pay very close attention to zoning issues near an airport, especially when a request is made for residential construction of any density.

The Concorde's inaugural flight was postponed by one day, but Kennedy Airport received dozens of calls on the day it was supposed to arrive, complaining of how noisy it was and how it should be banned. There are many other ways in which aviation is under incessant siege; when health care costs get mismanaged, the

nonflying public starts screaming for a luxury tax on airplanes to make up the difference. Government officials flogging pet aviation projects occasionally try to ram an expensive, superfluous program, such as microwave landing systems, down our throat.

Ours is an adversarial political system; competing goals and opinions slug it out to reach some workable compromise, and the loudest shouter stands to gain the most. So get out there and shout! Join or form groups directly countering a particular threat. Write letters to the politicians involved, outlining your views. Find public forums in which to state your views. Follow the rest of the tips presented in this section.

$ **Savings** Significant savings can be expected, but are difficult to quantify due to the vast differences from situation to situation. Your activism might help get fees and taxes thrown out or reduced, or your efforts might help prevent the closure of an inexpensive and convenient local airport, and so on.

199 Aviation's image projected positively

✻ **Action** Always be a good ambassador for aviation, which helps minimize antiaviation sentiment. Wonders can be accomplished by being reasonable, friendly, and persuasive, instead of being inconsiderate, or downright obnoxious. Make it your business to be aware of and observe night curfews and noise abatement procedures that are in place. Practice noise abatement procedures even when none are mandatory.

Avoid built-up areas at low level and expeditiously climb to altitude when flying over inhabited areas. *Never buzz anyone or anything*; there is no quicker way to set off an uproar by the general public. Be considerate of nonpilot's points of view, but also get your own position across, politely, persuasively, and persistently.

$ **Savings** Substantial savings are possible, achieved by reducing the chances for the implementation of expensive restrictions on flying.

General aviation interest groups

✻ **Action** Actively support or join organizations such as AOPA and EAA, representing general aviation's interests. General aviation interest groups are a powerful political force. Organizations such as AOPA and EAA have been very successful in countering expensive and unnecessary restrictions on general aviation by effectively presenting the objective facts from the flying community's point of view. But they can't accomplish the mission without your support, so do everything to provide that support.

These interest groups are also an excellent source for advice on handling challenges to aviation in your own community, and provide information on politicians friendly to aviation.

$ **Savings** Thousands of dollars might be saved during your flying career. Organizations such as AOPA and EAA leave thousands of dollars in your pocket during your flying career by getting current restrictive practices reduced, and by effectively countering moves to implement expensive and unnecessary fees, taxes, aircraft equipment requirements, medical requirements, and so on.

Support proaviation politicians

✻ **Action** Give proaviation politicians your vote and vocal support—if you can see your way clear of any disagreements that you might have with them on nonaviation issues. Many politicians at all levels are friendly to aviation; some of them are pilots. AOPA and EAA maintain a list of pilot-politicians; find ways to support them on their aviation positions.

$ **Savings** Substantial savings are possible, achieved by reducing the chances for the implementation of expensive restrictions on flying.

202 Support youth in aviation

✻ **Action** Do everything you can to get kids interested in flying, and to make it easier for them to fly. As flying expenses relentlessly rise and student starts fall, it is crucial to get as many people as possible into aviation.

A kid hooked on aviation is a participant or supporter for life. And we need all the new pilots we can get. You can do many things to get kids into aviation. Give rides regularly under the auspices of the EAA Young Eagles Program. Make aviation presentations in schools. Arrange aviation field trips to the local airport or an air route traffic control center or airshow, and so on. Join and become active in the Civil Air Patrol. Contribute to aviation scholarships.

$ **Savings** Savings are indirect, but significant. Every additional pilot chips away at the collectively high costs of flying. The more pilots fly, the less it will cost.

Resources

Aircraft type associations & aviation organizations

Aero Club of New England
P.O. Box 183
East Boston, MA 02128
617-973-7181

Aero Club of Pittsburgh
207 Dormont Village
2961 West Liberty Ave.
Pittsburgh, PA 15216
412-341-5090

Aero Medicos
2950 Mission Dr., Hangar 12
Solvang, CA 93463
805-688-0338

Aeronca Aviators Club
511 Terrace Lake Road
Columbus, IN 47201
812-342-6878

Aeronca Lover's Club
P.O. Box 3
401 lst St. East
Clark, SD 57225
605-532-3862

Aeronca Sedan Club
115 Wendy Court
Union City, CA 94587
415-471-5910

Aerostar Owners Assoc.
341 Albion St.
Denver, CO 80220
303-322-2376

AOPA
421 Aviation Way
Frederick, MD 21701
301-695-2000

Air Force Assoc.
1501 Lee Highway
Arlington, VA 22209-1198
703-247-5601

Alaska Airmen Assoc.
1515 East 13th Ave.
Anchorage, AK 99501
907-272-1251

American Bonanza Society
Mid-Continent Airport
P.O. Box 12888
Wichita, KS 67277
316-945-6913

American Helicopter Society
217 North Washington St.
Alexandria, VA 22314-2538
703-684-6777
703-739-9279 (fax)

American Maule Society
c/o Pick Point Air
P.O. Box 220
Mirror Lake, NH 03853
603-569-1338

American Navion Society
P.O. Box 1175
Municipal Airport
Banning, CA 92220
714-849-2213

American Tiger Club
and
National Bücker Club (aerobatic)
Route 1, P.O. Box 419
Moody, TX 76561
817-853-2008

American Yankee Assoc.
P.O. Box 1531
Cameron Park, CA 95682
916-676-4292

Antique Airplane Assoc.
Route 2, Box 172
Ottumwa, IA 5250
515-938-2773

Army Aviation Assoc. of America
49 Richmondville Ave.
Westport, CT 06880-2000
203-226-8184

Assoc. of Air Medical Services
South Raymond Ave., Suite 205
Pasadena, CA 91105
818-793-1232

Assoc. of Naval Aviation
5205 Leesburg Pike, Suite 200
Falls Church, VA 22041
703-998-7733

Aviation Crime
Prevention Institute
P.O. Box 3443
Frederick, MD 21705
800-654-5473
301-695-5444

Aviation Speakers Bureau
Suite 403
6475 E. Pacific Coast Hwy.
Long Beach, CA 90803-4296
800-247-1215

Balloon Assoc.
5630 South Washington Road
Lansing, MI 48911-4999
800-594-4634
517-882-8433

Balloon Fed. of America
P.O. Box 400
Indianola, IA 50125
515-961-8809

Beechcraft Duke Assoc.
P.O. Box 2599
Mansfield, OH 44906
419-755-1223
419-529-3822

Bellanca-Champion Club
P.O. Box 708
Brookfield, WI 53008-0708
414-784-4544

Bird Airplane Club
P.O. Box 328
Harvard, IL 60033
815-943-7205

Bücker Club
6438 West Millbrook Road
Remus, MI 49340
517-561-2393

Buckeye Pietenpol Assoc.
3 Shari Dr.
St. Louis, MO 63122-3335
314-966-0946

Cardinal Club
1701 St. Andrew Dr.
Lawrence, KS 66047
913-842-7016
913-842-1777 (fax)

Cessna Owner Organization
P.O. Box 337
Iola, WI 54945
800-331-0038
715-445-5000
715-445-4053 (fax)

Cessna Pilots Assoc.
Wichita Mid-Continent Airport
P.O. Box 12948
Wichita, KS 61211
316-946-4777

Cessna 150/152 Club
P.O. Box 15388
Durham, NC 27704
919-471-9492

Cherokee Pilots Assoc.
P.O. Box 7927
Tampa, FL 33673
813-935-7492

Civil Air Patrol
Maxwell AFB, AL 36112
205-293-6019

Commander Flying Assoc.
Suite E
899 W. Foothill Blvd.
Monrovia, CA 91016-1938
818-359-1040

Confederate Air Force
P.O. Box 62000
Midland, TX 79711-2000
915-563-1000

Continental Luscombe Assoc.
5736 Esmar Road
Ceres, CA 95307
209-537-9934

Corben Club
P.O. Box 127
Blakesburg, IA 52536
515-938-2773

Cub Club
6438 West Millbrook Road
Remus, MI 49340
517-561-2393

Culver Club
60 Skywood Way
Woodside, CA 94062
415-851-0204

Dart Club
3958 Washburn Dr.
Port Clinton, OH 43452
419-797-2434

DeHavilland Moth Club
1021 Serpentine Lane
Wyncote, PA 19095
215-635-7000

Ercoupe Owners Club
P.O. Box 15388
Durham, NC 27704
919-471-9492

EAA
Oshkosh, WI 54903-3086
414-426-4800
414-426-4828 (fax)

Fairchild Club
7645 Echo Point Road
Cannon Falls, MN 55009
507-263-2414

Fairchild Fan Club
P.O. Box 127
Blakesburg, IA 52536
515-938-2773

Fleet Club
4880 Duguid Road
Manlius, NY 13104
315-682-6380

Florida Aero Club
2808 North 34th Ave.
Hollywood, FL 33021
305-987-4266

Flying Apache Assoc.
6778 Skyline Dr.
Del Ray Beach, FL 33446
407-499-1115

Flying Doctors of America
1951 Airport Road
Atlanta, GA 30341
404-451-3068

Funk Aircraft Owners Assoc.
933 Dennstedt Place
El Cajon, CA 92020
619-466-1461

Great Lakes Club
P.O. Box 127
Blakesburg, IA 52536
515-938-2773

Hatz Club
P.O. Box 127
Blakesburg, IA 52536
515-938-2773

Heath Parasol
6431 Paulson Road
Winneconne, WI 54986
414-582-4454

Illinois Pilots Assoc.
P.O. Box 7367
Springfield, IL 62791
708-331-2117

Internat. Aerobatic Club
EAA Aviation Center
Oshkosh, WI 54903-3086
414-426-4800

Internat. Assoc. of
Natural Resource Pilots
200 Patrick St., S.W.
Vienna, VA 22180
703-560-1271

Internat. Bird Dog Assoc.
3939 San Pedro, N.E., Suite C-8
Albuquerque, NM 87110
506-884-4822

Internat. Cessna 120/140 Assoc.
P.O. Box 830092
Richardson, TX 75083-0092
612-652-2221

Internat. Cessna 170 Assoc.
P.O. Box 1667
Lebanon, MO 65536
417-532-4847

Internat. Comanche Society
P.O. Box 400
Grant, NE 69140
308-352-4275

Internat. Pietenpol Assoc.
P.O. Box 127
Blakesburg, IA 52536
515-938-2773

Internat. 180/185 Club
P.O. Box 222
Georgetown, TX 78626
512-863-3773

Internat. 195 Club
P.O. Box 737
Merced, CA 95340
209-722-6283

Luscombe Assoc.
6438 West Millbrook Road
Remus, MI 49340
517-561-2393

Meyers Aircraft Owners Assoc.
5852 Rogue Road
Yuba City, CA 95991
916-673-2724

Mooney Aircraft Pilots Assoc.
314 Stardust Dr.
San Antonio, TX 78228
512-434-5959

National Aeronca Assoc.
P.O. Box 2219
806 Lockport Road
Terre Haute, IN 47802
812-232-1491

National Aircraft Finance Assoc.
500 E St., S.W., Suite 930
Washington, DC 20024
202-554-5570

National Biplane Assoc.
Hangar 5, 4-J Aviation
Jones-Riverside Airport
Tulsa, OK 74132
918-299-2532

National Intercollegiate
Flying Assoc. (NIFA)
P.O. Box 3203
Delta State University
Cleveland, MS 38733
601-846-4205

National Ryan Club
4421 West 112th Terr.
Leawood, KS 66211
800-373-6202

National Stinson Club
14418 Skinner Road
Cypress, TX 77429
713-373-0418

National Waco Club
700 Hill Ave.
Hamilton, OH 45015
513-868-0084

National 210 Owners Assoc.
P.O. Box 1065
La Canada Flintridge, CA 91011
818-952-6212

North American Trainer Assoc.
25801 N.E. Hinness Road
Brush Prairie, WA 98606
206-256-0066
206-896-5398 (fax)

Northeast Stinson Flying Club
Brook Road
Simsbury, CT 06070
203-658-1566

OX-5 Aviation Pioneers
207 Dormont Village
2961 West Liberty Ave.
Pittsburgh, PA 15216
412-341-5650

Piper Owner Society
P.O. Box 337
Iola, WI 54945
800-331-0038
715-445-5000
715-445-4053 (fax)

Porterfield Airplane Club
1019 Hickory Road
Ocala, FL 32672
904-687-4859

Professional Aviation
Maintenance Assoc.
Suite 809
500 Northwest Plaza
St. Ann, MO 63074
313-739-2580
314-739-2039 (fax)

Rearwin Club
P.O. Box 127
Blakesburg, IA 52536
515-938-2773

Replica Fighters Assoc.
22451 David St.
Taylor, MI 48180
313-295-3926

Seabee Club Internat.
6761 N.W. 32nd Ave.
Fort Lauderdale, FL 33309
305-979-5470

Seaplane Pilots Assoc.
421 Aviation Way
Frederick, MD 21701
301-695-2083

Short Wing Piper Club
P.O. Box 66
Collinsville, OK 74021
918-371-9665

Soaring Society of America
P.O. Box E
Hobbs, NM 88241
505-392-1177

Southwest Stinson Club
14031 Elvita St.
Saratoga, CA 95070
408-867-7892

Staggerwing Club
1885 Millsboro Road
Mansfield, OH 44906
419-529-3822

Stearman Restorers Assoc.
823 Kingston Lane
Crystal Lake, IL 60014
815-459-6873

Straight Tail Cessna Club
2 Forest Lane
Gales, CT 06335

Super Cub Pilots Assoc.
P.O. Box 9823
Yakima, WA 98909
509-248-9491

Swift Museum Foundation
P.O. Box 644
Athens, TN 37303
615-745-9547

Taylorcraft Owners Club
12809 Greenbower Road
Alliance, OH 44601
216-823-9748

The Interstate Club
P.O. Box 127
Blakesburg, IA 52536
515-938-2773

The Ninety-Nines
Will Rogers World Airport
P.O. Box 59965
Oklahoma City, OK 73159
405-685-7969
405-685-7985 (fax)

The Skylarks of Southern California
15290 East Greenworth Road
La Mirada, CA 90638
213-943-1383

The Whirly-Girls
(Internat. Women Helicopter Pilots)
P.O. Box 58484
Houston, TX 77058-8484
713-474-3932

Travel Air Club
P.O. Box 127
Blakesburg, IA 52536
515-938-2773

Twin Bonanza Assoc.
19684 Lakeshore Dr.
Three Rivers, MI 49093
616-279-2540

U.S. Hang Gliding Assoc.
P.O. Box 8300
Colorado Springs, CO 80933
719-632-8300

U.S. Parachute Assoc.
1440 Duke St.
Alexandria, VA 22314
703-836-3495

U.S. Ultralight Assoc.
P.O. Box 557
Mount Airy, MD 21771
301-898-5000

University Aviation Assoc.
3410 Skyway Dr.
Opelika, AL 36801
205-844-2434
205/644-2432 (fax)

Vintage Sailplane Assoc.
Scott Airpark
Route 1, Box 239
Lovettsville, VA 22080
703-822-5504

Waco Historical Society
1013 Westgate Road
Troy, OH 45373
513-335-2621

West Coast Cessna 120/140 Club
P.O. Box 727
Roseburg, OR 97470-0151
503-672-5046

World Beechcraft Society
1436 Muirlands Dr.
La Jolla, CA 92037
619-459-5901

Zenair Assoc.
6438 West Millbrook Road
Remus, MI 49340
517-561-2393

States requiring aircraft registration fees

Registration fees are required in the following states. For specific and current information, contact the appropriate state aviation agencies.

Arizona
Connecticut
Hawaii
Idaho
Illinois
Indiana
Iowa
Maine
Massachusetts
Michigan
Minnesota
Mississippi
Montana
New Hampshire
New Mexico
North Dakota
Ohio
Oklahoma
Oregon
Rhode Island
South Dakota
Virginia
Washington
Wisconsin

State aeronautical agencies

Alabama Dept. of Aero.
555 South Ferry St., Suite 308
Montgomery, AL 36130-0101
205-242-4460

Alaska DOT
P.O. Box 196900
Anchorage, AK 99519-6900

Arizona Div. of Aero.
Room 426 M, Mail Drop 2191
2612 South 46th St.
Phoenix, AZ 85034
602-255-7691

Arkansas Dept. of Aero.
One Airport Dr., 3rd Floor
Little Rock, AR 72202
501-376-6781

California Div. of Aero.
P.O. Box 942874
Sacramento, CA 94274-0001
916-322-3090

Colorado Div. of Aviation
6848 South Revere Parkway, Suite 101
Englewood, CO 80112-6703
303-397-3039

Connecticut Bureau of Aero.
P.O. Drawer A
24 Wolcott Hill Road
Wethersfield, CT 06109
203-566-3933

Delaware Aero. Admin.
P.O. Box 778
Dover, DE 19903
302-739-3264

Florida Aviation Office
605 Suwannee St., M.S. 46
Tallahassee, FL 32399-0450
904-488-8444

Georgia Bureau of Aero.
2017 Flightway Dr.
Chamblee, GA 30341
404-986-1350

Guam Airport Authority
P.O. Box 8770
Tamuning, Guam 96911
(011) 671-646-0300

Hawaii Airports Div.
Honolulu, HI 96819
808-836-6432

Idaho Bureau of Aero.
3483 Rickenbacker St.
Boise, ID 83705
208-334-8775

Illinois Div. of Aero.
One Langhorne Bond Dr.
Springfield, IL 62707-8415
217-785-8500

Indiana Div. of Aero.
143 West Market St., Suite 300
Indianapolis, IN 46204
317-232-1477

Iowa Air and Transit Div.
Des Moines, IA 50321
515-281-4280

Kansas DOT
Topeka, KS 66612-1568
913-296-2553

Kentucky Office of Aero.
421 Ann St.
Frankfort, KY 40622
502-564-4480

Louisiana Aviation Div.
P.O. Box 94245
Baton Rouge, LA 70804-9245
504-379-1242

Maine Div. of Aero.
State House Station 16
Augusta, ME 04333
207-289-3185

Maryland State Aviation Admin.
P.O. Box 8766
Baltimore, MD 21240
301-859-7100

Massachusetts Aero. Comm.
10 Park Plaza, Room 6620
Boston, MA 02116-3966
617-973-7350

Michigan Bureau of Aero.
Terminal Bldg., 2nd Floor
Lansing, MI 48906
517-373-1834

Minnesota Aero. Office
State Transp. Bldg., Room 417
St. Paul, MN 55155
612-296-8046

Mississippi Office of Aero.
P.O. Box 5
Jackson, MS 39205
601-359-1270

Missouri DOT Aviation Sec.
P.O. Box 270
Jefferson City, MO 65102
914-751-2551

Montana Aero. Div.
P.O. Box 5178
Helena, MT 59604
406-444-2506

Nebraska Dept. of Aero.
P.O. Box 82088
Lincoln, NE 68501
402-471-2371

Nevada Dept. of Transp.
1263 South Stewart St.
Carson City, NV 89712
702-687-5653

New Hampshire Div. of Aero.
65 Airport Road
Concord, NH 03301-5298
603-271-2551

New Jersey Office of Aviation
1035 Parkway Ave. CN 600
Trenton, NJ 08625
609-530-2900

New Mexico Aviation Div.
P.O. Box 1149
Santa Fe, NM 87504-1149
505-827-0332

New York Aviation Div.
1220 Washington Ave.
Albany, NY 12232
518-457-2820

North Carolina Div. of Aviation
P.O. Box 25201
Raleigh, NC 27611
919-787-9616

North Dakota Aero. Comm.
P.O. Box 5020
Bismarck, ND 58502
701-224-2740

Ohio Dept. of Transp.
Bureau of Aviation
2829 West Dublin-Granville Road
Columbus, OH 43235
614-466-7120

Oklahoma Aero. Comm.
Room B-7, lst Floor
200 N.E. 21st St.
Oklahoma City, OK 73105
405-521-2377

Oregon Div. of Aero.
3040 25th St., S.E.
Salem, OR 97310
503-378-4880

Pennsylvania Bureau of Aviation
Transp. Bldg., Room 716
Harrisburg, PA 17120
717-783-2280

Puerto Rico Ports Authority
GPO Box 2829
San Juan, PR 00936-2829
809-723-2260

Rhode Island Div. of Airports
2000 Post Road
Warwick, RI 02886
401-737-4000

South Carolina Aero. Comm.
P.O. Box 280068
Columbia, SC 29228-0068
803-739-5400

South Dakota Dept. of Transp.
700 Broadway Ave., East
Pierre, SD 57501-2586
605-773-3265

Tennessee Office of Aero.
P.O. Box 17326
Nashville, TN 37217
615-741-3208

Texas Aviation Dept.
P.O. Box 12607
Austin, TX 78711
512-476-9262

Utah Aero. Ops. Div.
135 North 2400 West
Salt Lake City, UT 84116
801-328-2066

Vermont Agency of Transp.
133 State St.
Montpelier, VT 05602
802-828-2442

Virginia Dept. of Aviation
South Laburnum Ave.
Richmond, VA 23231-2422
804-786-1364

Washington Div. of Aero.
8600 Perimeter Road
Seattle, WA 98108
206-764-4131

West Virginia DOT
Bldg. 5, Room A109
Charleston, WV 25305
304-348-0444

Wisconsin Bureau of Aero.
P.O. Box 7914
Madison, WI 53707-7914
608-266-3351

Wyoming Aero. Comm.
Cheyenne, WY 82002-0090
307-777-7481

Index

Illustration page numbers are in **boldface** type.

About the author

Geza Szurovy has written three books for TAB/McGraw-Hill, Inc. *The Private Pilot's Guide to Renting & Flying Airplanes Worldwide* won the 1992 Aviation/Space Writers Association's Award of Excellence for outstanding achievement in aerospace communications in the books technical or training category.

Another book by Szurovy, *Fly for Less: Flying Clubs & Aircraft Partnerships*, received the 1993 Aviation/Space Writers Association's Award of Excellence for general aviation coverage in the books technical or training category.

Szurovy coauthored the third book, *Basic Aerobatics*, with aerobatic champion and U.S. Aerobatic Team member Mike Goulian.